Nearly a century ago two men — one in Scotland — reluctantly ..... miners and, in doing so, shaped the lives of all their joint Canadian descendants. In telling their story, the author of *Belly of Blackness*, who is one of those descendants, has created an exceptional tale of immigrant life in Canada in the early and mid-twentieth century, the crushing reality of the discrimination and privation, and the determination of those immigrants to be accepted and to prosper. This is an important story for every student of Canadian history.

Betty Keller
author of Better the Devil You Know

Belly of Blackness is a gripping tale brimming with heart, compassion and adventure. Elliot skilfully explores the rich history of family, delving deep into the dreams, heartbreaks and triumphs of his beautifully flawed and human characters.

Rebecca Hendry
author of Grace River

The story of coal miners and mining is vital to the history of Alberta. The miners and their families came from all around the world and contributed greatly to the social and economic make-up of our area. This book tells the story of two families and the hardships they faced. Thank you to Jim Elliot for sharing this story, which gives us some insight into the men who went daily into the darkness underground and the families who lived with the uncertainty of life in a mining town.

Belinda Crowson
Lethbridge Historical Society

Belly of Blackness

BELLY OF BLACKNESS
coal dust in my genes

Jim Elliot

Copyright © 2015 Jim Elliot

All rights reserved. No part of this publication may be reproduced, stored in a retrieval system or transmitted, in any form or by any means, without prior permission of the publisher or, in the case of photocopying or other reprographic copying, a licence from Access Copyright, the Canadian Copyright Licensing Agency, www.accesscopyright.ca, 1-800-893-5777, info@accesscopyright.ca.

Library and Archives Canada Cataloguing in Publication

Elliot, Jim, 1933–, author
    Belly of blackness : coal dust in my genes / Jim Elliot.

ISBN 978-1-926991-30-6 (pbk.)

    1. Elliot, Jim, 1933– —Family. 2. Coal miners—Alberta—Biography. 3. Coal miners—British Columbia—Biography. 4. Immigrants—Alberta—Biography. 5. Immigrants—British Columbia—Biography. 6. Alberta—Biography. 7. British Columbia—Biography. I. Title.

HD8039.M62C3 2015   331.7'62233409227123   C2014-907846-3

Cover image: © 2015 www.earldotter.com
Editor: Kyle Hawke
Proofreader: Bookmark: Editing & Indexing
Book designer: Omar Gallegos

Granville Island Publishing Ltd.
212 – 1656 Duranleau St.
Vancouver, BC, Canada V6H 3S4

604-688-0320 / 1-877-688-0320
info@granvilleislandpublishing.com
www.granvilleislandpublishing.com

First published in 2015
Printed in Canada on recycled paper

*To Mom and Dad
who lived the story*

Contents

Prologue ... ix

Chapter One ... 1
San Giovanni de Casarsa, Udine Province, Italy, 1907
A Rebel in the Family

Chapter Two ... 11
Glasgow, Scotland, late December 1907
Scots Wha Hae

Chapter Three ... 19
The Italian Immigrant

Chapter Four ... 33
Into the Belly of Blackness

Chapter Five ... 45
A Wedding in San Giovanni, June 1911

Chapter Six ... 65
In His Father's Footsteps

Chapter Seven ... 79
Discrimination, Frustration and Anger

Chapter Eight ... 89
Farewell to Scotland

Chapter Nine ... 111
The Alberta Coal Branch, January 1928

Chapter Ten ... 119
Dreams Unfulfilled

Chapter Eleven ... 145
Riding the Rails

Chapter Twelve
Independence 151

Chapter Thirteen
And the Two Shall Become One 157

Chapter Fourteen
Motherhood 171

Chapter Fifteen
And Baby Makes Three 185

Chapter Sixteen
High in the Mountains 197

Chapter Seventeen
No Relief 209

Chapter Eighteen
War Clouds 223

Chapter Nineteen
You're in the Army Now 237

Chapter Twenty
A Farmer He Is . . . Not! 251

Chapter Twenty-One
A Legacy of War 263

Chapter Twenty-Two
Out of the Pits 273

Epilogue 285

About the Author 288

Prologue

Though they had little else in common, both my father and my maternal grandfather mined coal in Alberta.

My grandfather, Umberto Francesco Linteris, was born into a wealthy family on August 3, 1885, and raised in San Giovanni de Casarsa, in the province of Udine in the Friulan region of northeast Italy. As he matured into manhood, Francesco, as he preferred to be called, rebelled against his father for supporting the oppressive Austrians who controlled that region of Italy. He decided to relinquish the comfortable life offered to him at home and seek the freedom and peace of mind that would come if he moved to oppression-free Canada. He emigrated in 1908 at the age of twenty-four.

My father, James 'Jim' Elliot, was born into a poor family on January 1, 1908, and was raised in Tillicoultry, Clackmannanshire — the smallest shire in Scotland. Tillicoultry, Gaelic for 'hill in the back land', is nestled at the foot of the Ochil Hills, which rise to the majestic mountain, Ben Cleuch. The hills would become a favourite hiking spot for Jim and the rest of his family.

For generations the eldest son of the eldest son has been named James, and my father's name was a continuation of this tradition. My paternal grandfather, Jimmy Elliot, was born into a wealthy Glasgow family, but had made decisions — including marrying a downstairs maid — that alienated him from his family and their wealth.

Jimmy and Francesco's stories are analogous to the Jonah myth in the Old Testament. When Jonah chose to ignore the life to which he had been called, he ended up in the belly of a whale — swimming in the blackness, praying for sunlight and dry land. The choices my grandfathers made led them into the black depths of a coal mine — enduring the constantly dripping water, the coal dust that drifted in the forced-air ventilation, the musty smells and the incessant noise of machinery digging and clawing its way through the body of coal.

My father followed his father into the mines when he was thirteen-and-a-half years old and the reality of the poverty in which his family lived shattered any dreams he had of becoming a doctor. For the next six years, his ambition was to leave the mines and break away from the predictable patterns of life that would squelch his spirit. He emigrated in 1928 to become a farmer in Canada but ended up in the coal mines of Alberta instead.

Both Francesco and Jim were part of the army of immigrants encouraged by colourful Canadian Pacific Railway (CPR) and Canadian National Railway (CNR) posters created on behalf of the Federal Ministry of Immigration and Colonization to entice people to Canada. The pictures showed open spaces and the words promised "Get Your Home in Canada" and "The Right Land for the Right Man." By the time Francesco arrived, immigrants made up 90 percent of the labour force in Alberta. The majority of these came from the British Isles (34 percent), Slovakia (23 percent) and Italy (15 percent), countries with a combined total land mass about the same as that of Alberta.

Many had come because of political, familial or economic oppression. Others had come for adventure or a temporary break from home. My maternal grandfather and my father had come to Canada to start over, and both discovered that this vast land of opportunity was not exactly as promised. Both knew what it was to be trapped in the belly of blackness.

Chapter One
San Giovanni de Casarsa
Udine Province, Italy, 1907
A Rebel in the Family

The young man slowed his horse to a walk as he entered the lavish courtyard of the Linteris estate on a fall morning. His buttoned brown coat marked him as an Austrian police officer and stood in stark contrast to the well-groomed flower garden that edged the stone wall, as much as his white pants and gleaming boots seemed out of place astride his horse.

He looked past the series of two-storey buildings that housed the family, toward the stable and the vineyard that rolled out across the undulating landscape. *Must be nice to have money*, he thought. He shook his head in disbelief that he had to participate in this sham. Why did these Italians not recognize their place? Why did money and influence have any bearing when it came to carrying out the law? He would rather have been in some more meaningful post in the service of Emperor Franz Joseph than this assignment in the province of Udine in the Friulan region of northern Italy — just because he could speak Italian. Although the native tongue of this region was Friulan, most of the people were taught Italian in school and he could make himself understood.

He dismounted gracefully in front of what appeared to be the main house and looped the reins over the railing beside the entrance. He lifted his hand to knock but stopped, instead adjusting his chinstrap to make sure his helmet was correctly in place and ironing the front of his buttoned brown coat flat with his hands. Satisfied, he gave the front door three sharp raps.

When it opened rapidly, he glanced at the document in his hand and looked up and down at the short middle-aged man in the doorway.

"I am here to speak to Umberto Linteris."

"I am Umberto. What do you want?"

He looked again at the paper.

"There must be some mistake. The man I am looking for is twenty-three."

"Let me see." Umberto scanned the document, emptied his lungs with a heavy sigh, shook his head and continued, "This is about my son, Francesco."

"Why do his papers show him as Umberto then?"

"We use his second name. I have an older nephew called Umberto and it was too confusing with too many with the same name. What has he done this time?"

"Your son made an anti-Austrian speech in the tavern."

"Again?" he sighed. "Why haven't you given him the castor oil treatment?"

The Austrian police kept a careful watch on young Friulan men, often forcing them to swallow a large dose of castor oil if they thought they'd broken some law.

"From what I have been told," the officer snapped, "it probably would not stop him. Anyway, my superior told me to remind you that it would not be good for *your* name if we arrested him."

Umberto knew that he was able to maintain his position in the community by acquiescing to the Austrian authority and did not want to upset the status quo.

"Be assured that I will see to my son myself." His stern blue eyes met those of the officer as he offered a stiff hand. "Thank you."

The young man ignored the gesture. He clicked his heels together, turned, strode back to his horse, mounted it and rode off, a stiff poker fused to his spine. Umberto watched him leave, then spun around, facing into the house, his back to the open door.

"Arturo!" Umberto bellowed into the house. "Where is that scoundrel brother of yours?"

"He's in the stable," Arturo said, rushing into the room, "tending to his horse."

"Get him now!"

Arturo ran from the house. Soon after, Francesco sauntered from the barn toward the house.

"Get a move on," Umberto ordered. "*I need to talk to you!*" The hot blood of his rage turned his face a deep red as he glared at his approaching son. "What are you doing to me?"

"What do you mean?"

"A policeman just arrived with more complaints about your loud mouth."

"You don't need to worry," Francesco said curtly. "This is my problem. It's not about you."

"*Anything* that involves *my* family is about *me*. You want to destroy us."

"I was only talking . . ."

"Be quiet and listen! I've been very patient with you and given you a good chance to make something of yourself. And this is all the thanks I get — a son who wants to go to jail."

"Papa, I don't like the way things are . . ."

"You don't like . . . ? I am tired of you spouting off your radical ideas about the government and encouraging the workers to ask for more. You are lucky I am well-respected or you would be in a cell right now. We still have a good name but when I hear that 'Umberto Francesco Linteris has done this,' or 'Umberto Francesco Linteris has said that,' or 'Umberto Francesco Linteris is a playboy,' I don't want to know you!"

"I should be able to think whatever . . ."

"Don't think! Just settle down and keep your mouth shut or get out of my life!"

When the conversation finally concluded, Francesco's ears burned as if they had been cuffed. With his father's rage fuelling his own, he retreated silently to the barn, and quickly saddled his horse, Libertât. He had carefully chosen her name five years earlier when his father had given the silvery-white yearling to him as a reward for hard work and a controlled tongue. He remembered thinking, *She will carry me to my freedom when I need to get away from here. The older I get, the harder it is to keep quiet.* Libertât had been a faithful companion to him and seemed to sense his need for speed as she galloped effortlessly across the hills on those days when Francesco just couldn't bear to be around his father.

"Let's go girl," he said lovingly, patting her neck, and they headed toward the village.

The grapes had all been harvested and the resident hired help, Carlo and his family, could easily handle the few chores without Francesco's help today. Riding along the lane past the twisted grapevines, he foresaw them budding the next spring. *Another crop and one more year of the same monotonous routine.* Would this be his future? It was family tradition that when one of the children married, a new two-storey apartment was added to the family complex and the newlyweds would live in it and raise their children there as the previous generations had done.

He remembered his father's ultimatum when he had returned from Munich: "You are welcome to live here and raise your family here, but it will be with my rules and you will keep your place. I want no more trouble with the law." As he rode on, he thought back to when he was seventeen. At that time, following a particularly intense argument with his father, he had run off to Germany to seek like-minded people who would be sympathetic about the situation in Friulan.

Leaving, though, had offered no respite. Munich seemed an escape when he was seventeen, but the others his age cared nothing for politics — or serious conversation of any kind. Nights at parties trying to engage them had left him disappointed, and days working in the beer stein factory kept him from working towards anything better. In the end, he'd decided he could make more of a difference back at home. The only positive thing about the trip had been experiencing the peace and beauty of the mountains as the train pulled slowly over the Brenner Pass.

Now, as he rode toward the village, Sister Mary Immaculata came to mind. She had been a bit of a stealthy rebel in her history lessons. He could hear her voice now as she told of the formation of the Friulan region following Napoleon's defeat in 1814. How it became one of the Austrian duchies and remained part of Austria even when the southern regions formed the United Kingdom of Italy under King Victor Emmanuel II in 1861. And how now Austria was allied with Hungary under a dual monarchy to increase its military strength and furnish the resources necessary to maintain secure control of the Friulan region. Francesco remembered her emphasizing how important this was as two rail routes ran north from Trieste — one to western Europe and the other to eastern countries. One of the routes ran through San Giovanni and, although this link helped the local economy, most Friulans wanted freedom from the oppressive regime to the north.

Sister Immaculata loved the region and encouraged her pupils to appreciate who they were and how they got there. Sometimes she sounded like a poet describing how the people of Friulan had created a cultural quilt with threads of the invading groups running through both language and food customs. The climate produced some of the best wine grapes in Italy and provided ideal conditions for growing mulberry plants, which led to a large silk industry.

She encouraged her pupils to be proud people, recognizing their uniqueness in so many ways — especially in their surnames, which usually ended with a consonant — not a vowel as in the rest of the Italy.

Francesco had admired her, especially as her teachings seemed to be the opposite of what Father Giuseppe said in his homilies. The sister's enthusiasm had stirred in Francesco a pride in his roots, which led to a deep dissatisfaction with his father's acquiescence to the ruling Austrians.

His stomach sloshed bile up into his throat as he rocked back and forth in the saddle, reflecting on what it would cost him to settle down in San Giovanni de Casarsa. He could not talk to his father without running the risk of getting his ears boxed, so he usually held his tongue until he was with his friends in the tavern, where he expounded at great length about the oppressive Austrians, the rights of local workers, and how the church was a pawn in the hands of the oppressors. The Pope had issued guidelines for the local priests, urging them not to speak against the regime, and Umberto supported that stance — another sore point that could not be included in any discussions at home. All of this had been stirring in Francesco for some time and his thoughts often moved into alternatives. *I will not spend the rest of my life on the family estate — nor anywhere else in this country under the present regime.*

Francesco had a well-deserved reputation as a ladies' man, but lately he had begun courting Angelina Clarot, a beautiful woman six months older than him who had attended the same school. The petite curly-haired brunette lived with her father, Valentino, and her two older brothers, Luigi and Benedetto, in a comfortable two-storey house on their small vineyard. Her mother had died giving birth to her, and Valentino, who had never remarried, had raised his three children alone while earning a modest living from his family-run vineyard. Angelina cared for the home, cooking and cleaning. Her older brothers pampered her and kept a protective eye on her as she matured. Nevertheless, she and Francesco found time to be alone and nurture a relationship that was becoming serious.

Francesco was torn, as he realized he was falling in love with Angelina and wanted to spend the rest of his life with her. But he would not do so the traditional way. She deserved better and he would make sure she got it.

Over the next few weeks, as Francesco drank wine and talked with his friends at the *osterie*, the conversations inevitably turned to discussion about leaving the country. They all had seen posters provided by the Canadian Pacific Railway and the Dominion Coal Company encouraging men to take advantage of the opportunities in Canada, where there was plenty of work, especially in the coal mines. The colourful drawings showed endless fields of grain, The Italian government, in response to the tough economic times, also displayed posters encouraging people to move to America.

There had been a steady flow of discouraged Italians emigrating to the Americas over the past thirty years, and some who had returned home on visits told stories of endless opportunities in the new land. These accounts and the colourful travel posters carried more weight in the mind of young idealist Francesco than did the reports in the Milan newspapers telling of low wages, poor working conditions and the hard labour borne by many Italians in the new world. He and his friends just wrote them off as coming from a few disgruntled people. In fact, the more he and his friends discussed it, the more convinced Francesco became that he must go to that part of America they called Canada.

Francesco pondered the possibilities as he rode home on a December night in 1907. When he arrived at the barn, he dismounted, lit the oil lamp, and led Libertât to her stall, where he took his time grooming her. He would have to leave her if he was to realize his dreams.

"I will miss you, my faithful friend."

He rubbed her neck one more time. *If it is so hard to leave my horse, what will it be like leaving Angelina?* He knew he needed to talk to her.

Francesco moved in and out of sleep all night, his mind like a big pot of polenta being relentlessly stirred, ideas rising to the surface and popping back into the constantly churning mush. In the morning, his thoughts still in turmoil, he kept out of sight in the barn fussing with some tack, delaying his departure until Angelina's father and brothers would be outside doing chores.

Later that morning as he approached Angelina's home, he was relieved to see that there was no sign of the Clarot men near the house.

"Francesco, what are you doing here?" asked Angelina, startled.

"I need to talk to you," he said quietly.

"What's wrong? You don't look so good."

"I'm fine . . . but Angelina, I can't take it anymore."

"What? Me?"

"Not you. Living here. Living at home. My father . . ."

"We can go away . . ."

"That's what I want to tell you. But I have to go first."

"What?"

"Angelina, my *preciôs*, I want to go to Canada to work and make enough money to come back and marry you."

"I don't understand why . . ."

"I have thought it out and made up my mind, so listen to me. I'm not getting married now and spending the rest of my life living under the thumb of my father and raising my children to be just like everyone else. I want something better for you and me and I have to go away to make it happen."

"No! We can go together." She threw her arms around his neck. "I won't let you go."

Her tears blurring her vision, she silently clung to Francesco, acknowledging what they both knew — *it is the man who decides* — she wept.

"Angelina, I love you and it will be good for us one day. But I must do this now."

After a lingering kiss, he released himself from her arms and slowly walked to his horse. Angelina stood alone, her eyes filled with tears, watching her blurred love ride off.

A week later, when the Linteris family gathered to celebrate Christmas, Francesco's 20-year-old cousin, Roberto Francescutti, announced that he and his friend Garcchin, who was also a stonemason, were going to America in the spring.

"Garcchin has a friend working in a coal mine in the mountains of Canada — a place called Michel — and he thinks we can get jobs as masons in the big city of Fernie. There are a lot of Italians."

"You're too young to go," interrupted Roberto's mother, Donatella.

"I'm three years older than Francesco was when he went to Germany," Roberto replied.

"Don't bring that up," said Umberto. "It is hard on the mother when a son leaves. Families need to stay together like we always have." He got up to hide his discomfort and picked up a bottle of wine. "Who wants more of my beautiful Tocai?"

Later that evening, Francesco invited Roberto to accompany him while he checked on the horses.

"Do you think Garcchin's friend could get me a job?" asked Francesco.

"But you're not a stonemason," said Roberto.

"No! In the mine."

"Well, it sounds like there's lots of work there," replied Roberto.

"I think I'll take my chances," Francesco said. "When do you go?"

When Roberto had finished detailing his plans, Francesco asked why they were sailing from Le Havre, in France.

"It takes less time," replied Roberto. "I was looking at the posters. If we leave from France, we get to America in just over a week. If we go from Italy it takes up to twenty days."

"It sounds very good. I'd like to go with you."

The next day, when all their guests had gone, Francesco told his parents that he had made up his mind to go with Roberto.

"You're still too young," his mother pleaded.

"I am three years older than Roberto, and Aunt Donatella is letting him go."

"Reluctantly. But at least he has Garcchin to look out for him," she said. "They would not even be working in the same town as you."

"I did all right in Munich and I was only seventeen then. I can handle myself."

"I told you then, and I will tell you now: You can live here, you can work here, you can raise a family here, but it will be with my rules and . . ."

"That's why I need to go, Papa," Francesco butted in.

"Then you go on your own. I will not help you leave or return." Umberto rose and walked toward the door. "I am done with you."

The next day, Francesco rode over to the steamship office in Pordenone, where he stopped to peruse the colourful posters one more time. Once he was inside the office, a travel agent helped locate Michel, British Columbia, on the map that the CPR had circulated with its posters, and Francesco happily purchased a ticket for a sailing from Le Havre.

When Francesco arrived at Angelina's door a few days later, she opened the door and looked at him without inviting him in.

"Francesco, I just don't . . ."

"Angelina, it's all set!" he blurted. "I'm going to Canada with Roberto. We're leaving . . ."

"What about me?" she stopped him mid-sentence, glancing nervously behind her. "I'm expected to wait?" She began to cry. "I love you and I want to believe you, but it is so hard."

"I told you that I will come back when I have . . ."

"You scoundrel!" Valentino appeared behind his daughter, putting one hand on her shoulder and pointing an accusing finger at Francesco. "Look what you are doing to my Angelina. You fill her with promises and then break her heart just like you did to all the rest of your women."

"But I love her and I am . . ."

"You show your love by running away?" He took a step toward Francesco.

"I'm not running away," Francesco pleaded, offering his open hands. "I have to go to Canada to better myself."

"Nothing is going to make you better. You're a libertine." Valentino shook his head and moved back beside the silent Angelina in the doorway.

"You deserve your reputation."

"But I love Angelina and I *will* come back for her." Francesco was close to tears.

"It's a good thing I sent the boys off this morning or you'd get more than words from them." He took a step toward Francesco again. "Now get out and don't torture my daughter anymore."

"Papa," Angelina burst out, following her father after the retreating Francesco, "don't send him away." The stern face that had greeted her fiancé was now softened by tears. "I need to talk to him."

"He's no good for you!"

"At least let us talk, Papa. I believe him."

"If you *are* lying," Valentino said, again pointing a finger in Francesco's face, "you'll regret it for the rest of your life."

He stomped back into the house.

"Angelina," Francesco reached for her hands, "I love you and I promise on my grandmother's grave that I will return. Then we will get married and move to Canada."

He explained about the ticket and his travel plans and promised to see her again before he left.

"Go, if you must, my love." She stepped back from Francesco. "Please hurry back."

Chapter Two
Glasgow, Scotland
Late December 1907
Scots Wha Hae

On a cold winter afternoon, Martha Elliot reached over the sink and wiped the misty window with her warm hand. She peered out to see if her husband, Jimmy, was on his way home. As she stared blankly at the dreariness of Glasgow's Maryhill district, she felt that the bleakness had somehow seeped inside her. The steady winter drizzle had long since washed any colour from 'the green' below, with its sod clumps scattered helter-skelter in the soggy clay, the unpainted picket fences matching the drabness of the row-house walls, and the smoke from the washhouse chimneys adding to the greyness. She shook her head as she gazed at her neighbour's dripping wash, precariously supported by a slanting pole that barely kept it out of the muck. *Poor lass. She gets even more discouraged than I do. And there's nothing in the sky to promise a change so her wash will get wetter before it gets dry.* Martha's line, strung across the end of the kitchen, was never empty of daily-washed nappies, except on sunny days when they could be refreshed outdoors.

With another baby due any day now, she was tired much of the time and the weariness covered her in a blanket of despair. She looked at her children and was grateful that they seemed to understand her need for a peaceful day. At thirty, Martha already had three children. Seven-year-old Meg, just back from school, was practicing her numbers on the slate her aunt had given her. Cissy, aged four, cradled her new teddy bear in her tiny arms as she watched her older sister carefully, wishing she could print like her. Dolly, aged two, had not yet wakened from her afternoon nap and was buried in the pile of blankets on the bed in the alcove. *What bonnie bairns, but it's no' been a guid year. This would have been our worst Christmas ever if Lizzie had not come by with . . . what did she say . . . "A few extras for the weans."*

She wiped the window for one more look — no one in sight. She stood, gently rubbing the stump that hung from her right shoulder. An accident in the Glasgow rubber factory when she was twelve had mangled her arm so badly that it had to be amputated just above the elbow. It ached sporadically, especially in the cold of winter and, even after all these years, she sometimes had the sensation of feeling in the missing fingers. *He's stopped at the pub again. Telling more stories.* She grunted and slowly shook her head. *To think I used to like hearing all about the way things were in days gone by.* Although surrounded by many people, Martha had limited her human contacts to a few of the other wives as they hung clothes in the green. She wanted out of it all, to move away from the city to a smaller, quieter, cleaner, healthier place.

The ache in her arm was matched by that in her heart. She turned from the window and went to the stove to stir the pot of gray stovies and prevent this mess of potatoes, onions and sausages from sticking to the bottom. *Stovies again. But what can you do if all you have is tatties, sausages and onions? How much better it would be if Jimmy got a decent job!*

There were shops close by selling vegetables and meat and bread, but Martha was not in a position to purchase much and did not leave home very often. Thankfully, the 'onion Johnny', his bicycle draped with braids of freshly-picked French onions, had come by that week and she had been able to get a good supply.

It is always good to see that black beret coming up the street. Those Johnnies are not afraid of work. So far away from their homes in France, riding their bicycles throughout the winter. Jimmy could do with some of that ambition.

Moving the pot back, she opened the stove lid and bent to pick up another lump of coal from the meagre pile in the bin. Coal was expensive and Jimmy often didn't work enough to keep the bin full. Bending was not easy these days and straightening up even harder. With a bit of a wince, she dropped the coal into the fire, closed the lid on the cast-iron range and moved her pot to just the right spot to keep it simmering.

Och, it's time to make the tea. I'll probably drink it alone while he drinks that demon drink. Now, simmering along with the stovies, she waddled back to the sink and filled the kettle from the swan-necked tap, then peered through the foggy glass again, hoping he would be home before she reached the boil. *He always finds enough money for a pint or two, even with just his part–time painting job. And me with three bairns tae feed. He's got to find more work. I'm so tired of never having enough to eat.*

Returning the kettle to the hottest part of the stove, she then emptied the teapot, dumping the spent leaves from breakfast on top of the potato peelings in the bucket beside the stove. Before going to bed, she would scoop some coal dust from the bin and mix it with the burned cinders and whatever else she had collected into the bucket that day, place them on the fire, and hope that it would smoulder all night to combat the chill in the cold, damp house.

She crumpled into a chair and stretched out with her feet on the pouffe that she had improvised from a box and a cushion. Clasping her stump with her left arm, she rested it on her bulging abdomen and closed her eyes. *I'll just sit a wee minute. How I would love to go to sleep. I would have been better off staying in Sheffield as a maid. Jimmy and his promises!* "We'll go back to Glasgow. There'll be more work and my family is there." *He said it didn't matter that I was a downstairs maid and he was the eldest son of a wealthy Glasgow family, or that I didn't have much education and he was well-learned, especially in history. His family . . .*

She was jolted back from her daydream when Dolly awoke and her sharp cries announced her need for attention. Martha pushed herself up and out of the chair, moved the simmering pot of stovies farther back on the stove so they wouldn't burn, pulled a dry nappy from the line and exchanged it for the wet one that was making Dolly so uncomfortable.

"Meg, come and look after your sister while I make the tea."

"I'll do it," offered Cissy.

Cissy went to the baby and began to sing one of the songs she had learned in Sunday school for the Christmas program.

"Away in a manger, no crib for a bed . . ."

Dolly smiled at her sister and appeared to enjoy the performance.

"Meg, will you set out the plates?" Martha asked. "I'll get the supper out."

"What about faither?"

"I'll put his by on the hob. Get at it."

When she and the girls had eaten, Martha returned to her chair and surveyed her brood. They were all sitting on her bed singing. *How will we manage in this crowded place with another mouth to feed? We canna live here much longer. Too many weans are dying of sickness in this dirty, crowded city. It's not right for the bairns.* She knew that the overcrowding was pushing the authorities to develop a better health and sanitation system, but it would be years before many of the residents reaped the rewards of this program, and Martha did not plan to wait there until it happened.

Two years earlier, when they had decided to move to the Maryhill district of Glasgow, 'Scotland's Venice', it had been with the hope that Jimmy's family would give them a hand. But it was not to be. His parents were unhappy that he had married beneath him and accused Martha of looking for a way out of the slums. Jimmy had stood up to them and they in turn had disowned him. His older sister, Liz, and her husband, John Cullen, did not agree with the parents' decision and often came by with groceries for the larder and gifts for the children. For Jimmy, it was not easy to accept the fact that he was now a labourer who needed to look for work.

Maryhill seemed like a good place to be. It had grown quickly in the late 18th century when the canals had been built to connect the Forth and Clyde rivers, and many industries moved in to take advantage of access to the seas on either side of Scotland. But, like many fast-growing communities, it experienced a myriad of social problems. Alcohol abuse was rampant — at one point there was one pub for every fifty-nine inhabitants. This had led to the formation of the first temperance union in Great Britain in 1824. Since there was more profit in building pubs than there was in building houses, good accommodation was difficult to find. Jimmy and Martha had no choice but to move into a small two-room second-storey flat with an alcove off each room. The larger alcove in the main room was for a bed, while that in the small bedroom was enclosed in a cupboard to be used as a bed-closet or 'hole-in-the-wall' bed. Recognizing the need for adequate ventilation, Glasgow city council had passed an act in 1900 that forced landlords to open the front of the bed recess from floor to ceiling for at least three-quarters of its width. The space below the beds was available for storage.

In the main room, the stove and a coal bunker that could hold about two hundred pounds of coal — as if they could ever afford to fill it — were on the same wall as the entrance door, with the table and four wooden chairs on the opposite wall. Between the table and the coal bunker on the side wall stood a long chest of drawers for dishes and cutlery and other kitchen utensils, with a shelf above it for larger items. A doorway beside the chest led to the children's bedroom. An alcove, recessed into the fourth wall, contained the bed where Martha and Jimmy slept. The children slept in the small bedroom with the alcove bed piled high with blankets to help keep out the cold.

While Martha waited at home, Jimmy Elliot was sitting with two of his cronies, Alex and Willie, in the Whitehouse pub, carefully nursing his ale. The Whitehouse had been built about 150 years earlier when work had begun on the canals. Situated at lock 21 on the canal, the highest point between Glasgow and Edinburgh, it stood like a large white boot on the banks of the canal — the long foot butted against the two-storey main building. The white stone gave the stark building its name. Over the years, it had witnessed many tall tales and many more ales. Tonight, the topic was how well the Glasgow Celtics — those blasted Irish Catholics! — were doing and what a poor season their beloved Rangers were having and how it looked like even Hibernian was not going to be able to knock off the Celtics. As was usually the case when Alex was there, the conversation shifted to politics.

"CB really pulls for the little fellow," Alex declared.

"Aye, it's hard when you have to deal with those Sassenach Conservatives. They block his every move to reform things," Willie replied.

CB, or Prime Minister Henry Campbell-Bannerman, had been born and raised in Glasgow and had served the Stirling Burghs as MP for almost forty years.

"I tore this out of the *Herald* the other day," Jimmy said, retrieving a folded piece of newspaper from his pocket. "'Personally, I am an immense believer in bed, in constantly keeping horizontal: the heart and everything else goes slower, and the whole system is refreshed.' That's our CB talking. A man after my ain hairt."

"Yes, and Mattie will have your heed as well as your hairt if you don't get more work," said Willie.

Jimmy stood. He pushed his chair back, took a deep breath and stabbed his finger into the space between him and Willie.

"Dinnae change the subject, mon. We were talking aboot the great Scotsman running this country."

"Will we see you on Hogmanay, Jimmy?" asked Alex softly in an attempt to divert the conversation.

"I may not get out first-footin'. Mattie's due to have the bairn any day noo. I am hoping she will wait until the new year."

"Ha'e ye got a name for the wee lassie?" teased Willie.

"It had better be a laddie," replied Jimmy with some indignation. "I'm needing one to carry on the Elliot name and the tradition of naming the eldest son James." He sat down again and fingered his empty glass. "You ken that song, 'My name is little Jock Elliot, and wha daur meddle wi' me?'"

Alex and Willie both grinned. They knew Jimmy's practice of throwing one last story into the night before reluctantly heading home for another tongue-lashing from Mattie. And the family name meant the tale would be one about the *reivers* from whom it descended. The men were well aware that the Elliot clan had been very prominent raiders along the English border in the 13th century.

"Aye," said Willie, raising his glass, "we learned that as children."

"Well, did ye know that Jock Elliot almost caused the death of Mary, Queen of Scots?"

"I didna' know that," said Alex, laughing. "What did he dae, gie her a fright by just looking at her?"

"Dinna be daft, mon. I am serious. She didna hae much to do wi' her second husband, Darnley, and was sweet on Jimmy Hepburn, the Earl of Bothwell. Back in 1566, they planned to ha'e a wee tryst at the Hermitage."

Jimmy looked from his empty glass to the fading light outside the window and was tempted to order another ale when he remembered his empty pocket.

"Ye ken," he continued, "that Bothwell was a recklessly brave, intelligent man who tried to impress the queen by cleaning out some of the reivers in the area. Well, he made the mistake of rounding up a number of Elliots and placed them under heavy guard in his castle."

"Surely they would have put up a fight," Alex egged him on.

"Our clan had just successfully ended a long feud with the Scotts. They were feeling good and he caught them unawares. Then Bothwell made a big mistake. He decided he would go out on one more expedition before his rendezvous with the queen."

"He really wanted to impress her, I guess," interjected Willie, winking at Alex.

"Aye. The stupid man," Jimmy replied, really into his story now. "He went out alone and, not far from the castle, he came upon the notorious Little Jock Elliot. Bothwell shot Jock out of the saddle and then foolishly dismounted to see if the prone man was fully dead. Little Jock was all over him in an instant and stabbed him three times wi' his dirk. Bothwell's men, standing guard outside of the castle, heard the ruckus and rushed into the forest. They found their bleeding chief and rushed back, ignoring the wounded Elliot. They arrived to find that the Elliots had overpowered their guards and were now in charge. They made a wee deal with him that allowed him to stay in his own castle while they went awa' hame safely."

Jimmy sighed. He stood and began to walk toward the door.

"I thought this was about Queen Mary," said Alex.

Jimmy paused and turned around.

"The Queen got lost trying to find Bothwell and she wandered into a wet marsh. She got a terrible cold and almost died."

"So what happened to Jock?"

"He died a little later from his wounds but not before he had left his mark and his reputation. Later, they wrote that song about him." Jimmy looked out the window. "Och! It's getting fair dark, I'm in for it noo. See you the morrow."

"Dinnae fall into the canal, laddie," Willie said, raising his glass to his departing friend.

Jimmy knew well the route along the canal and down the slope to his street. Many times he had left the Whitehouse to head home somewhat inebriated. Going home was always a little easier — downhill all the way. It was downhill all the way for everyone.

When he had first moved to Maryhill, he would often marvel as he stood on the banks of the Kelvin River, watching it flow under the arches of the viaduct while a ship passed overhead on its way from Edinburgh to Glasgow. He also knew that the canal had claimed a few imbibers over the years and so he was careful to stay on the footpath.

As he wearily climbed the worn brick stairway up to his flat, he braced himself for the welcome he usually received when late home from the pub. Quietly opening the door, he peered into the dimly lit room and saw Meg sitting at the table practicing her writing by the oil lamp. Cissy was playing with Dolly on the bed, Martha dozed in a chair, and a plate of warm stovies sat grayly on the hob above the stove. He closed the door and went for his dinner. Meg looked up at the sound of the closing door.

"Hello, faither," she said. "I'm working at my lessons."

"Haud your wheesht, lassie!" Jimmy urgently whispered. "You'll wake your mother."

The warning came a little too late.

"Jimmy, this has got to stop!"

It was amazing how quickly Martha could move from sleep to all-out rant.

"Did you find any work today?" she demanded.

"No, I was at . . ."

"The Whitehouse," she snapped as she rose and went to the stove, picked up the plate of stovies and slammed it on the table. "Here, you'd better eat something."

"Mattie, I tried. There just isn't anything," he said quietly.

He jabbed a fork into the mass in front of him, searching for a piece of sausage.

"The bairn's due any time noo, and ye ha'e to be here for the girls and not out drinking with your heathen friends."

Jimmy didn't like being jobless. By the time he faced Martha, he was more exhausted than he might have been had he worked all day. He ate his supper without speaking, nodding agreement to whatever his wife said. There was no way he could be dominant with this woman. His underemployment and not infrequent intemperance undermined any possibility of that ever happening. Martha really was, as many women were in those days, 'the angel of the home'. And the angel wasn't finished with him just yet.

"And furthermore, I'll no raise any more bairns here. You've got to get better work away from this dirty city. Something that will keep food on the table."

"It's not that easy, Mattie. I've been to a number of places. There's no steady work for a painter or paper hanger," he said with some exasperation.

Martha poured them each a cup of tea and the conversation followed a familiar pattern, ending, as usual, with Jimmy finishing his dinner in silence with a lukewarm cup of tea, while Mattie got the children into their beds.

When Hogmanay arrived on that last day of December in 1907, Mattie knew by the stirrings inside her body that the next child would soon make an appearance.

"Jimmy, you had better go over to Nellie's and tell her the pushin's have started and I need her."

Mattie and Jimmy got their first-foot at 11 p.m. New Year's Day when a dark-haired son was born. And since the best first-foot should be a dark-haired man, maybe their luck was changing.

"Mattie, you've done it, lass! You've given me a son and wee Jim will carry on the name of the eldest of the eldest. I think I will ha'e a wee dram to celebrate."

"Make it a very wee one. You've got to care for the bairns the next couple of days. I am very tired and you have to here and be sober."

Chapter Three
The Italian Immigrant

Francesco, Roberto and Garcchin left Le Havre for Canada on March 21, 1908, aboard the steamship *La Gascoigne*. The voyage across the Atlantic took ten tedious and uncomfortable days in the cramped quarters below decks that they shared with the other third-class passengers, mostly other Italians. The German and French passengers largely had cabins above deck. The dullness of the March days and the lack of light and fresh air in the dormitory-like cabins might have depressed Francesco if he had not seen the whole thing as a great adventure.

On March 31, Francesco joined the others on the deck as they approached the Statue of Liberty. Once the ship had anchored, the passengers were ferried to the dock at Ellis Island. Since all of the first and second-class passengers disembarked first, it seemed an eternity before Francesco's group finally boarded the little boat and joined the long line-up awaiting processing by immigration officials.

The Ellis Island immigration depot was a processing centre for third-class ship passengers arriving in New York Harbor (first and second-class passengers were processed on board ship prior to disembarking). The new arrivals were ferried from their transatlantic vessels to Ellis Island where they were guided in groups into registration areas in the Great Hall, a room 200 feet long and 100 feet wide, filled with chain-link fence corridors to direct the people back and forth in one long snaking line to the processing area. The three young Friulans stayed together and endured the process of having a tag attached to their coats showing the manifest number of *La Gascoigne*. Interpreters assisted the inspectors, and a registry clerk recorded their responses. Because all three of them had papers showing their final destination in Canada, the initial interrogation only lasted two to three minutes — but it seemed like forever, as they had to respond to a list of questions designed to discover any signs that the immigrant might pose a problem to the country.

Upon completing the registration process, where each of them had the letter 'N' written next to their name — denoting the more-respected north Italian — they were ushered into a room where medical officers observed their movements in 'the six-second exam', checking the scalp, face, hands, neck, gait and general condition. Next came a more formal inspection where the examiner took a buttonhook, a metal instrument used to button gloves, and pulled each eyelid back to look for signs of trachoma. Francesco noticed that some people had their outer garments marked with white chalk.

"What are those markings?" he asked the translator.

"Oh," the man replied, "H is for heart problems, Pg for pregnancy, E for eye problems, and L for lameness. You're done here. Next, you will get a short intelligence test to weed out idiots, imbeciles or morons and other mentally deficient persons."

The young men encountered no problems in this stage and then faced one more in-depth interview with more probing questions about reasons for emigrating and their plans. This whole process took between three and five hours to complete and finally they were given beds for the night in a large dormitory filled with cots. Following breakfast the next morning, they rode the ferry to New Jersey and boarded the train heading for Montreal. Once there, they transferred to a Canadian Pacific Railway immigrant train bound for the west coast.

The luxury Francesco had known travelling by train in Europe was nowhere to be found on this leg of his journey. The rugged, drafty cars were equipped with hard seats and filled to capacity with a variety of people speaking a jumble of languages. Francesco and his companions found seats together at the opposite end of the car from the smoking coal stove, and had to bundle up at times during the cold, dreary journey across the vast country. They were only able to join the crowd at the warm end when someone there found it too hot and sought a cooler part of the car. This movement back and forth helped develop a sense of camaraderie among the passengers as they mingled and shared the experience, usually relying on a third person to help translate.

There were as many different cooking smells as there were people. However, the trio stuck mostly to the bread, cheese and milk they purchased from stores near the railway stations across the country. When they reached Toronto, some of the immigrants disembarked and Francesco asked one of the passengers if they were almost at Michel, British Columbia.

The man smiled as he answered.

"No, it will be many more days before you get off this train."

It took almost two days to traverse the rough terrain of the Canadian Shield with its monotonous trees, rocks and lakes. Then, at last, the land began to level out as the train followed a winding river valley to the next major stop, Winnipeg. After a three-hour wait while the crew changed, the young Italians began the last leg of their trip — two long days in which the only breaks in the bleak landscape were scrub trees and giant red grain elevators. Finally, late one morning, they reached Lethbridge, and when they got off the train to pick up supplies, they saw them: *mountains*.

"We're getting closer," Francesco declared. "Michel is in the mountains."

Francesco sat back in his seat as they pulled slowly out of the station, his gray eyes focused on his reflection superimposed on the verdant backdrop. He liked what he saw — well-tanned olive skin with a nose that did not overstate his Italian roots.

It seemed to take forever as the train wound down into a river valley and then slowly climbed out of it, following the rolling contours of the foothills as the train crept into the mountains. It slowed even more as they passed over the largest rockslide he had ever seen — the railroad tracks were right on the rocks. The only evidence of habitation was a little red building with a white sign that read 'Frank'.[1]

A couple of hours later, the conductor entered the car. He stopped beside the young Friulans, nodded and announced, "Michel."

Roberto smiled.

"Michel," he slowly repeated. "We are here."

They grabbed their trunks, stepped onto the wooden platform and set their belongings down while they took in their surroundings. They were in the middle of a small town that sat in a narrow valley between some majestic snow-capped mountains. Almost obscuring the mountains on one side, gray-brown smoke rose up from behind a long brick wall. Closer in,

---

1   On April 29, 1903, at 4:10 a.m., 82 million tonnes of limestone crashed from the summit of Turtle Mountain and buried a portion of the sleeping town of Frank. This bustling town was home to approximately 600 people. Of these, roughly 100 individuals lived in the path of the slide. An estimated seventy people died. The primary cause of the Frank Slide was the mountain's unstable structure, but underground coal mining, water action in summit cracks and severe weather conditions may have contributed to the disaster. The buried section of railway was rebuilt three weeks after the slide, and a road through the slide was completed in 1906.

a long row of single-storey unpainted wooden houses stood beside the rail line, stretching out toward some larger buildings, one of which was painted pink with 'HOTEL' in large black letters on one wall.

"Garcchin!"

The three turned as one to see a grinning young man rushing toward them.

"Pietro! What a relief to see you."

Garcchin introduced his friend to the others.

"Come, I have a wagon waiting," Pietro beckoned the three of them to follow.

"What's the smoke?" Francesco asked.

"Oh, it's just the coke ovens. They never stop."

"I don't know what coke ovens are."

As they rode to the hotel, Pietro told the newcomers all about how coal produced in the Crowsnest Pass mines was high-quality bituminous. When burned, it produced little ash and was very good for generating steam in locomotives and power plants, and the Canadian Pacific Railway used it extensively. However, one of the most important uses of coal from the Crowsnest was in the manufacture of coke, which is used in the smelting of copper and iron ores in southern British Columbia, Montana, Idaho and Washington state.

The train pulled out for Vancouver and revealed a row of two-storey houses on the other side of the tracks.

"Why don't the people paint their houses?" asked Roberto. "Everything is so colourful back home."

"The smoke from the ovens destroys paint quickly, so they save money this way. They are nicer inside," Pietro said. "Many of the people who live in those houses all take in boarders, but they're all full now."

"Where will we stay?" asked Francesco.

"I got a job for you in the mine," Pietro assured him, "and arranged for you to stay in the bunkhouse with me. The mine runs the place like a hotel for single men. You get your meals and washing done. Garcchin, I have arranged for you and your friend to stay in the hotel for a couple of days. I will go with you to Fernie on my off day."

"Is Fernie bigger than this?" asked Garcchin.

"Oh yes," Pietro laughed. "It is the commercial centre for this whole area and is growing. You will have a good chance finding work as masons."

"Thank you for doing all of that. You must really know your way around here," said Francesco.

"I do. You'll be able to get all the clothes you need for the mine in Fernie."

Once they had deposited their trunks, Pietro showed the new arrivals around the little town and began to educate them on some basics of living in Canada. He told them how the 'Englishers' ran the mine and all the main businesses, and that they sometimes treated immigrants from other countries with contempt.

"We don't use our old country names here. I am now Pete."

He went on to suggest that Francesco might be better to anglicize his name to Frank. Roberto would be Bob and Gary would be perfect for Garcchin.

"I think the English don't like to write and they want short names," he chuckled.

On Saturday, they all took the train to Fernie. While there, Pietro introduced Roberto and Garcchin to a friend who had agreed to put them up for a few days and help them find jobs as stonemasons. He took Frank to the mercantile and assisted him in making the necessary purchases for work in the mine — boots, cap for his lamp, heavy pants and shirt and long johns — taking delight in demonstrating the amazing buttoned trapdoor in the back of them. When they got back to Michel, Pete suggested that Frank scratch his name on to his lunch pail, as they all looked alike.

As Frank carefully worked a small chisel over the shiny surface, he thought, *This is the first time I have done my name in English. Frank Linteris.*

Prior to starting his first shift, Frank was given a carbide lamp and, with Pete translating, learned how to use it. First, he had to put calcium carbide in the lower chamber of the lamp and then fill the upper chamber with water. A notched lever on the side controlled the rate of dripping water and thus the amount of it used to control the rate at which the water drips into the generator. Adjusting the rate of flow determined the amount of acetylene gas that the reaction produced, and thus the size of the flame. A small flint wheel built into the lamp provided the spark to ignite the gas. A reflector behind the little gas jet helped produce a surprisingly bright, wide beam. There were some electric lights in the main tunnel but, while he was working, his only source of light would be that lamp. The lamp man then handed Frank a handful of brass discs, all with the number 122 on them.

He joined the miners as they piled into a small train of wooden coal cars. Though he considered himself an adventurer, he was frightened when the cars began to roll down the slope into the blackness, their only connection with the outside world being the cable that attached the cars to a winch. He hunched down in the cold air and wondered if he would be warm enough, even though he was wearing his new long johns.

After what seemed a very long time, the cars came to a halt. The men climbed out and began walking along a dark tunnel lit only by the lamps on their bobbing heads. When they came to a line of empty cars on the track, Frank was assigned a work area next to Pete, who explained that he was to shovel the loose coal into the empty cars. It was the toughest job in the mine and the one reserved for new and apprenticing miners.

"Let me see one of those discs you got from the lamp man," Pete said.

Frank reached into a pocket and pulled out a brass disc with a number on it.

"Hang one of these on each car you fill. You get paid by the amount you load."

It took a while for Frank to coordinate the movement of his arms and his head so that his work area was properly illuminated. The shovelling was backbreaking and relentless, and Frank tried to keep up to the rhythm of the others he could hear along the work face. The cold fresh air pumped through the mine to clear out any gas also brought a variety of smells with it.

One smell he recognized had him a bit puzzled, as it triggered memories of the barn back home — manure. Then he heard a new, yet familiar sound — a horse approaching. *I must be going crazy in this blackness. I saw some horses pulling cars on top, but horses down here?* His sanity was affirmed when a pony came into the range of his lamp pulling an empty coal car with a young boy riding on the front of it. He could not understand what the lad said, but while the boy unhooked the empty car and hooked up to the cars he and Pete had filled in order to take them out to the main tunnel, Frank took a moment to nuzzle the horse's neck. *Beautiful.*

Seeing the horse brought a bit of sanity to Frank's new environment, but as he worked, the dust continually floated around him. He had to stop from time to time to cough and spit out the muck that accumulated in his mouth and throat. How different it was from back home where all he had to do when he was cleaning out the dusty barn was step outdoors into the brilliant sunshine and inhale the pure Italian air. Now he felt trapped in the blackness with no way to escape the dust.

Fortunately, an older miner came by and offered him a pinch of snuff, explaining that it would help keep his mouth moist so that he could frequently spit out the dust instead of swallowing it. Frank stuck some in his mouth between his lower lip and bottom teeth as instructed and immediately felt that he would have to get rid of the vile-tasting stuff. However, he persevered, determined not to give in, and soon was breathing a bit easier. Every part of him ached and it was a relief when the men broke off work for their sandwiches. Frank opened his new lunch pail and used the first few mouthfuls of tea to rinse and spit. There was little conversation during lunch but a great deal of coughing — a sound that would become familiar.

That night at the supper table, he ate silently, too tired to talk. The others laughed and told him that they had all been through it and he would be fine in a few weeks. He had never worked like this and wasn't sure he could keep it up.

However, he soon got into a routine and his muscles stopped aching so much and started developing. He was determined to fit in and become a Canadian as soon as possible; learning the language would be the beginning. He bought an Italian-English dictionary and the first of many Westerns to help him. Every night he would sit on his bed in the bunkhouse reading, translating each word, slowly learning the new language and a great deal about cowboy life. Sometimes it was hard to concentrate with the noise of the conversations from some of the other eleven men who shared the space. Each bed had a small table beside it with a coal oil lamp on it, some hooks for clothes on the wall and room for a trunk under the bed.

When he had been in Michel about two months, he received a disturbing letter from Angelina. She explained that she had thought she might be pregnant before he had left, but because she wasn't sure, she had said nothing. Now, she was sure. Her letter made it clear how she felt.

Can you come home now? I am frightened and don't know what to do.

It must have been that night I told her . . . Why wasn't she more careful . . . ? Once again his mind turned into bubbling polenta. Did he really want to go back now that he was free? He looked intently at the picture he had pinned to the wall above his bed. *I am a Linteris; I will do what is right . . .* But he had just starting putting a little money away and knew it would take some time to save enough for his fare back to Italy and two more fares to return to Canada.

He sat down and wrote a careful reply.

You will have to be patient. I am doing the best I can, working very hard, but it is not easy to save money quickly.

He went on to ask what her father and brothers thought about the pregnancy and if his parents knew. Not sure what else to say he simply closed with:

Each day when I am in the blackness, you are my light. I think of you, and the day we will be together. Ti ami, I love you.

Frank worked diligently and did not complain about the coal dust that caked onto his clothes as it mixed with the continual dampness in the dripping mine. By the end of each week, when he took his work pants back to the boarding house to be washed, they were so stiff that it seemed as though they might be able to stand up by themselves.

Because he wanted to save money, he limited his social activities to an odd night out with some friends at the local hotel. They talked of the similarity between life back home under the Austrians and the present situation where racial slurs were part of their everyday life.

He quickly learned that 'wop' meant scoundrel, rogue or thug. 'Papist' was a belittlement of Roman Catholics. Frank was angered, not just because of the names he was called, but also because all of the advantages went to the Englishers. He also realized that Michel, where he couldn't even use his own name, was not like San Giovanni, where his father's position had given him some freedoms to voice his opinion without fear of reprisal. Here, he was on his own and could not complain lest he lose his job, but he was determined that nothing would stop him from reaching his goal. He and the other Italian immigrants had brought a character trait with them, *arrangiarsi*, which means making do with what you are given and seeing the acceptance of adverse conditions as evidence of strength of character and moral fibre. It had sustained his family for generations and now would get him through the discrimination and whatever else he had to face in Canada.

Francesco stuffed his anger deeply inside and was even more resolute in his decision to become a Canadian as soon as possible. He pored over his novels for hours at a time, making notes of the new words he encountered. Some of the other boarders teased him about his method of learning, suggesting that he would end up more of a cowboy than a miner. Nevertheless, they also encouraged him and helped when he had difficulty conversing in English around the meal table.

One day in June, Pete told Frank that he had just received a letter from Gary in Fernie saying that he and Bob were not getting much work and were going to see if things might be better in Lethbridge.

"Is Lethbridge bigger than Fernie?" Frank asked.

"Yes. Don't you ever write to Bob?"

"No. We were never that close back home," Frank replied. "Anyway, I have my hands full writing to my sweetheart and my mother."

During the hot summer evenings, Frank usually went for a walk after supper to be alone with his thoughts before going to his bunk to write home or plod his way through a few more pages of his current western. On Saturdays, he usually joined Pete and his friends at the hotel for a couple of beers and an opportunity to develop his ear for English.

On August 1, the conversations around the small tables were interrupted when someone from the telegraph office rushed in with the news that a huge fire had just wiped out Fernie. This stirred up memories for some miners of the devastating fire that had burned Michel to the ground in 1902. The next day, Sunday, as clouds of smoke billowed in the southwest, a CPR train backed into Michel from that direction. The crew announced that they had not been able to get through — the fire was moving rapidly up the valley, advancing ominously toward Natal and Michel. Since the CPR trains could not proceed west of Michel, they were available to shuttle the fleeing townsfolk to Coleman, where they could stay until the danger was over.

On Sunday afternoon, Frank packed his trunk and got on the first available train east. He was not going to risk losing his possessions — he had no real attachment to Michel. When he arrived in Coleman, he rushed to the hotel and was fortunate to find a room to share with three other men from Michel. After a good meal in a café, he set out to explore Coleman. It was much bigger than Michel and although there were also rows of coke ovens, the valley was more expansive, like the Tagliamento, and the smoke had more room to dissipate. He could see snow-capped peaks in the distance.

Frank came upon a building with a small sign over the door, 'Italian Hall', but because it was Sunday, it was closed as were all the businesses in town. He would ask about it.

When he got back to his room, he wrote to Angelina, telling her that he had moved. The next evening, he discovered a couple of men who had come from Friulan and asked them about the Coleman mine.

"I think they're always hiring," said Emile, formerly Emilio. "It's a good place to work."

"The mine is quite new," said Steve, previously Stefano. "They have a washhouse with lockers for your work clothes. You can make lots of money here if you work hard. Are you going to stay in Coleman?"

"For now," Frank replied.

"Well," Steve said, "if you decide to get a job here, there's an empty room where we live."

"I'll go to the mine tomorrow," said Frank. "I saw an Italian hall here."

"Yah," said Steve, "we get together to sing, dance and play cards regularly. The women always make great food. It's good to have a place where we can relax without any disparagement."

"Is it like this all over in Canada? Are Italians second-class everywhere?" Frank asked.

"Yes," replied Emile, "but at least we don't have it as bad as the bohunks — the nickname they give the Ukrainians and other central Europeans. They have their own section just west of town and we don't mix much with them."

Frank was successful in getting a job. The next day when Steve got off work, he met Frank at the hotel and together they carried his trunk to his new living quarters. On arriving at the boarding house, Steve introduced Frank to Mrs. Panatti, the Italian immigrant who owned the house.

"*Entrate. Entrate,*" the rotund, matronly woman said as she wiped her hands on her apron and reached out to greet Francesco. "I'm just making supper. Call me *Madre*."

Indeed, she did remind him of his mother and he hoped she could cook like her.

"Steve, show him where he sleeps."

At supper, Frank met the other boarder, Tony, originally Antonio. This was so much homier than the bunkhouse in Michel and Frank sensed that he was going to like it here.

At supper one night, the topic of the new opera house came up.

"It should be finished this fall," Emile said.

"An opera house?" said Frank. "I love opera."

When he had turned fifteen, his father and his Uncle Carlo had included him in their annual trip to Venice for a few days of the opera season. Francesco told the others around the table that it had been a rich experience to be exposed to the works of Donizetti, Rossini, Pucinni, and the beloved Verdi, with all the opulence that La Fenice Opera House offered.

"Do you know why they call it La Fenice?" he asked.

They all shook their heads.

"La Fenice means 'the phoenix'," he continued. "It's had three fires since it was first built and they keep raising it out of the ashes, just like the phoenix in Greek mythology. But my best time was when we went to

Milan in 1906 to attend a performance of Verdi's *Nabucco* at La Scala. I was twenty-one years old. I am still stirred as I think about it."

"It won't be like that here, Frank," interrupted Steve. "In Canada, they build places for entertainment and call it an opera house. I'm afraid you won't hear any Verdi."

Following supper, Frank went to his room to compose another of his regular letters to Angelina. He also wrote to his mother, who was upset that she was going to be a grandmother *and that the parents of the child were not married.* In every letter, she asked why he could not come home and do the honourable thing. She said that his father agreed, although he would not send any money to buy Francesco's passage home.

After Francesco's departure for Canada, Valentino had begged Angelina to forget the Linteris boy. As a wise and loving father, he knew that this man would not make a good husband for his precious daughter. In addition, Valentino was well aware that, should Angelina marry Francesco and emigrate to Canada, he would likely never see her again. In an effort to persuade her to stay, he began to build an extension on his house for Angelina as was the custom when children set out on their own.

However, within a couple of months it had become obvious to her family that Angelina was pregnant and that the father of the child had to be that scoundrel in Canada. Now, Valentino was even more determined to protect and care for his daughter. With the help of his sons, he soon finished the new quarters for his daughter and her coming child. As the pregnancy progressed, he was in a constant state of agitation. Twenty-three years earlier, when his wife was pregnant with their third child, they had all been so happy, had celebrated his birthday and then almost as though she were giving him a gift, she went into labour. It was a very difficult birth, and as the baby girl took her first breath, the mother took her last and was gone. Valentino tried to hide the pain of his memory and his anxiety for Angelina's well-being from her and his sons, but they knew the cause of his turmoil.

Angelina felt smothered by her father's control over her life during her pregnancy. She went to mass on Sundays where Father Giuseppe always had a kind word for her. Sometimes, she was allowed to accompany her father to the market, but Valentino was embarrassed when he ran into people he knew and rushed the shopping to take his daughter back to the privacy of their home. She knew that tongues wagged in the village, and even at church she could feel the darts of the stares fired her way, but she felt so alone and longed for contact with her friends.

"You have brought enough disgrace to this family," her father would say when she suggested she might go for a walk into the village.

So she busied herself looking after the household, avoided arguments with her brothers and her father and controlled the tears when they were around. Alone in her bedroom, she would read the letters from Francesco as she caressed her abdomen and dreamed of the family that was yet to be. She cherished his letters and read them over and over. How romantic he was — a side of him she could not share with her family for they would not understand. She knew he was working hard and saving as fast as he could, but over and over she wished the waiting was over. *How long is it going to take for Francesco to get here. He must try before the baby comes.* But this was not to be.

When Angelina gave birth to a baby girl on November 8, 1908, eight months to the day from Frank's departure, she had already determined the name — Maria, after Francesco's grandmother and Catherina after her grandmother. When the midwife had left, her father and brothers came into the room to see the baby.

"Maria Catherina," Angelina proclaimed, "this is your grandfather and these are your uncles."

"That is a very beautiful name," remarked Valentino, taking in the sight of his dark-haired granddaughter. "She will be strong with the name of the Holy Mother and your grandmother. Your mother would be so proud."

Angelina did not enlighten him on the meaning behind her choice of names.

From the outset, the child was spoiled. A grateful grandfather and two proud uncles helped make the beautiful Maria feel special and adored. They also continued to encourage Angelina to forget Francesco and make a life for herself in San Giovanni.

When Frank received the news that Angelina had given birth to a girl, he smiled especially at the name. *It was clever of her to use Catherina as the second name. I don't think she knows that it is also my mother's name.* However, when he read that the two of them were living in her new place on her father's farm, he began to worry that she just might get too comfortable in that situation and not want to move to Canada.

As the months slowly passed, he struggled to push back the discouragement that filled him, forcing himself to be content with a few pictures of his daughter and the letters telling of her rapid growth and her developing precocious personality. Reading about how much her grandfather and uncles doted on Maria filled him with a resentment toward them that built with each letter. Meanwhile, his savings grew more slowly than he wanted and he could do nothing except work hard and continue to save.

His mother wrote often, telling him that his father had decided that they would not see the baby until Frank returned and was properly married.

"I talked with Father Giuseppe and he will marry you in the church. Your father has made arrangements with him but doesn't want you to know," she wrote. "He expects you to live here and work when you get home. We will have our granddaughter with us then."

Frank had to word his letters home very carefully, making sure he did not make reference to some of the subjects in his mother's correspondence, knowing his father always read all of the mail.

In March 1911, the miners in the Crowsnest Pass went on strike for better working conditions. The Coleman miners, even though they had a better situation than most, decided to support their colleagues and left the job. Frank had not quite reached his financial goal, but as talk of a long strike persisted, he decided to return to Italy and marry Angelina. He could stay there for the summer and harvest season, work to make some money and then he and his family could return to Canada. In preparation for his trip, he visited Bill Donaldson, the owner of the Merchant Taylor shop in Coleman, and had a suit made so that he might return looking as dapper as possible. In early May 1911, Frank Linteris boarded the train in Coleman and began the lengthy, tiresome trip home to Italy.

Chapter Four
Into the Belly of Blackness

Through that winter and into the summer, Jimmy Elliot persevered with job-hunting but the work he found was never steady. He returned to tramping through Glasgow looking for something better. That too was in vain. Martha was relentless in her nagging and the pattern was as predictable as the tides at the mouth of the Clyde in Greenock. The arguments persisted until one evening as summer was moving into fall.

"Jimmy," cried Martha, "you're a lazy heathen possessed by the de'il who needs to pray more and drink less."

"I'll no take any more of your tongue, lass," Jimmy said, rising from his chair.

"You'll take what I give you until you take your responsibilities as a husband and father," she replied, waving a wooden spoon at him.

"You dinnae need to threaten me with your spurtle. There's no work here and all your harping won't make any. I'm going to Tillicoultry in the morning."

"Tillicoultry?"

"Aye. There's got to be plenty of jobs in the mills."

"Just like that, you're awa?"

"I've been thinking about it for a time. I can stay with Robert Wilson. Pack me a jelly piece for the morning."

"We don't have money for the train."

"I can walk. It's only thirty miles. I've walked more than that around Glasgow."

Martha was silent for a moment. Then, with a hint of encouragement she offered, "There are mines as well as the mills there. And they're paying good wages."

"I'll no' go into the mines, lassie. No Elliot has stooped that low and I won't be the first. It's not fit for man nor beast."

Early the next morning, he stuffed the jelly pieces and a flask of tea into a small bag, along with an extra shirt, and set off for Tillicoultry, a town nestled at the foot of the Ochil Hills near the Devon River. As the sun warmed, so did he, yet he plodded on.

I'll show her that I'm not lazy, he thought. *I'll find a way to make it better for my family. Och, my feet are getting hot, but I'll no' give in and I won't lower myself to ride in a farm cart.*

When he reached Stirling and saw the castle above him with Wallace's Monument proudly proclaiming a fine day in Scottish history, he knew he was getting closer to the end of the trek. The next few miles through Fishcross and across the Devon into Tillicoultry were a bit easier. Late in the day, he finally arrived at the home of his friend Robert Wilson.

Following a good night's sleep, a substantial breakfast and some encouraging words from Robert, Jimmy began his job search with renewed energy. However, as he went from mill to mill to mill, it soon became clear that things were not much better in Tillicoultry than they had been in Glasgow.

"Have you been out to the mines?" asked Robert over the welcome dinner that followed Jim's second fruitless day of footslogging.

"I'm no' going into the pits!" Jimmy said emphatically. "None of my family has ever worked in that slave trade and I'm not going to be the first."

"With winter coming on, they're hiring a number of folk," said Robert. "It will be better than the dole with your brood."

"It's the last thing I'll do," replied Jimmy despairingly. "I'll keep on looking. Someone must need a painter or the like."

Day followed discouraging day and Jimmy's spirit began to languish as the beautiful blooms of summer withered in the dreary fall. He walked over to Menstrie then to Dollar and down to Alloa, but the results were the same. No one needed what he had to offer.

One day, with his heart as empty as the coal bin at home, he trudged down Alva Burn to where it met the Devon River. He looked across at the tipple of the Alloa Coal Company mine and decided to give it a try. He glanced down at his dusty, scuffed boots — hard to distinguish from the road they trod upon — and walked.

As he walked, his thoughts turned to his family. How he missed them, especially his firstborn son — but he knew he couldn't return to Glasgow and face Martha without finding work.

Mattie's no a bad woman. She's a guid mother . . . a guid wife . . . it was her strength that I saw when I first met her. It's no' been easy for her with all the bairns. She doesna mean ill of me . . . but I wish she'd hold back on her tongue now and then.

He reached the mine office and reluctantly signed on for a job loading cars at the coalface. While he was relieved to finally have a job that would enable him to move his family to Tillicoultry, he was ashamed and dejected that he was had been forced to do the work that had always been the lot of the lowest of the low.

It took two Saturday paycheques before he had enough money — with Robert's help — to rent a council house at 79 Alexandra Street, in the Devonside area just across the river from Tillicoultry, and move his family into the council house that would be their home for the rest of their lives.

Jimmy hated the dark, damp, cold mine: the constant creaking of the timbers, the throbbing roar of the dust-spewing coal-cutting machine, the water constantly dripping from the ceiling, the mucky film that grew on his clothes when dust and water met, the coughing men, the scurrying rats and the vile smell that followed him home. Even the gentle breeze of the relentlessly pumped air was a discomfort as it chilled his skin and refused to allow his clothes to dry. The work was exhausting. He monotonously shovelled coal into a hutch, dragged it into the cross-cut, dumped it into a coal car and went back to do it again, hour after endless hour every day except Sunday. Although the job provided a bit more for his family, he was not used to hard work and continued to find excuses to miss shifts. Therefore, once again, he and the family had to rely on financial assistance from the council. This was hard on Jimmy, as he was a proud man and did not like being on the dole, but he pushed his pride aside and accepted whatever was available from the council. Not only did they live in subsidized housing, but were also recipients of clothing made from heavy worsted wool for long-term usage and boots with tacks pounded into the soles so the leather would not wear out. It was impossible for those on the dole to walk quietly on the cobblestone streets and Jimmy felt ashamed to be wearing them when everyone knew that they were part of the outfit the council gave to folk on the dole. There was no way to walk quietly as the tacks in the soles announced your poverty long before anyone saw you coming.

Weeks moved into months and months into years and the persistent patterns continued. Neither he nor Martha liked what he was doing, but there was no other way for them, especially with the rhythm of their life adding a new child every two years — Mary in 1910, Jock in 1912, and Bill in 1914. Each day blurred into the next with constant arguments about his lack of ambition and the resulting scarcity at home.

Martha took solace in her weekly trek with the children to the Baptist church while Jimmy continued to find comfort in a glass of ale with his cronies. Through all of this, they agreed on one thing — none of their children would ever go into the mines. As he watched his firstborn and namesake grow, he was determined that wee Jim, as they called him to save confusion, would never go underground.

One evening in early 1915 as the girls cleared up after supper and Martha and Jimmy were having a cup of tea, she sternly took Jimmy to task.

"Now that the war is on, and there is need for more coal for the factories, why aren't you working as much as the other men?"

"I am doing what I can. You know how much the dust bothers my eczema. I can only manage a few days at a time," Jimmy replied.

"And all the dust and smoke in the pub is good for you? You're a lazy lout."

Jimmy knew it was futile to continue to argue with his wife, she would never understand how hard it was for him to be in the mine. He looked over at wee Jim, now seven years old, and smiled. He was a bright conscientious boy, always helping his mother when she needed an errand run and who had recently taken up a paper route to bring in a little extra for the family.

"There's where the ambition is in this house," said Martha as she followed Jimmy's gaze to wee Jim. "He knows how to be responsible and doesn't belly-ache about the work, even though some of his grumpy customers expect a lot from a young lad."

Jim had learned how efficient it was to heave the paper from the road onto the front stoop of most houses, but his progress was slowed at a couple of places where he had to walk to the front door in response to complaints from grumps. Recently he had found a pair of old wheels, picked up some scrap wood from behind the mill and, with help from his father, crafted his pushcart. The pushcart not only made it easier to get around with the papers, but the wheels were wobbly and squeaked on the axle, masking some of the sound his boots made on the stones.

The sign in the Co-op window, 'Delivery Boy Wanted', caught young Jim Elliot's attention, and the eight-year-old immediately saw an opportunity to earn more money. It was 1916, the world was still at war, and at home, his parents continued to war over his father's inconsistent work habits. Jim, as the eldest son, had his special place in the family instilled in him from his birth and, as he grew older, was determined to be more industrious than his father.

He parked his pushcart where he could keep an eye on the newspapers and entered the store. Walking up to the counter, he peered over the top.

"Hello, Jim, what does your mother need today?" inquired the man in the long white apron.

"I'm not here for my mother, Mr. Gordon." Jim quickly replied. He turned and pointed to the window. "I saw your sign and I'd like that job delivering groceries."

Mr. Gordon smiled as he looked down at the diminutive boy in his council-supplied clothes — a worsted jacket and pants and a pair of 'tackedy' boots. He knew Jim and his family well, as Mattie always brought one or two of the children with her when she shopped.

"Does your mother know you are here?" he asked gently.

"No, but she'll say it's all right."

"How old are you now, lad?" Mr. Gordon inquired.

Standing as tall as he could and looking up into his questioner's eyes, he replied," I'm eight and I'm good in school. I can read and do arithmetic and I've been delivering the *Tillicoultry News* for a year."

"Well, the groceries can be heavy, you know?"

"I'm strong and I made a pushcart to carry my papers. It will be good for groceries, too," Jim replied eagerly.

"You've got a healthy spirit, and I like your mettle. If your mother says yes, then I'll give you a try."

"Thank you, Mr. Gordon," Jim beamed. "I will do a good job and make you proud."

He turned, headed out to finish his route and hurry home. *Every Saturday morning. And I'll get a shilling for it. Mother will be pleased.*

Near the end of his route, Jim began to anticipate supper and wondered what meagre meal would be waiting. *Probably soup or boiled neaps. Maybe mother bought some sausages and made stovies. I wish I were rich and could buy meat and she could make pies . . .*

Jim relished those rare occasions when there had been enough money to buy beef and his mother had made a meat pie. With no oven at home, one of the children would take it to the baker's for cooking. He loved to accompany Meg when she picked up the pie as he always got to carry it home, savouring the delicious smell.

His six-year-old sister, Mary, was waiting for him two blocks from home "to help my big brother" with the last part of his route.

"I've a new job, Mary," Jim said excitedly. "I'm going to deliver groceries for the Co-op."

"I'll tell Mother. She'll be happy," said Mary, spinning around to begin skipping away.

"Hold on!" Jim hollered. "I want to tell her myself."

Mary turned and ran back to the cart.

"Come on now, Mary. You do that side of the street and we'll be home in no time."

"Okay. I bet I can do my side faster than you."

He liked being the big brother to wee Mary as she was more playful than his older sisters who had more responsibilities around the house, cleaning, cooking, washing and caring for their two youngest brothers who, at four and two years of age, were more of a nuisance for Jim, especially when his pals were around.

As usual, Jim let Mary win the paper race and as she raced ahead to the front door, he called after her.

"Not a word to mother about the Co-op."

Martha, in the middle of supper preparation, was pleased when she heard the news that gushed forth from her son.

"Good lad! Will Gordon is a fine man. Make sure you stick in with it," she said encouragingly, putting down her stirring spoon and ruffling Jim's hair. "Not like your father. With a war on and all the young men off to fight and not enough miners, he still doesn't stick with it."

"If I was older, I'd go and fight the Hun!" Jim stood at attention and saluted.

"Just stick in with the groceries, lad," his mother said as she turned back to the stove.

Jim grew with the job and handled the responsibilities admirably as he delivered groceries all over Tillicoultry. Proudly and dutifully, he gave his mother the shilling each week from the Co-op. It wasn't much, but she was a frugal buyer and now and then was able to save enough for a decent meat pie. He kept back the few bonus pennies he received from grateful customers who appreciated his diligence. He didn't have a sweet tooth himself but liked to surprise his younger brothers with Jujubes or aromatics or Jelly Babies once in a while. But he saved the Dr. Jim Thin Toffee as a reward for Mary, who often accompanied him on Saturdays. Even with this generosity, he was still able to buy tobacco and at eight years of age started smoking whenever he was away from the house.

Jim was the favoured child in the midst of the six siblings. For Martha, he was becoming what she hoped: a hard worker who emulated her attitude and stick-to-itiveness. He was industrious, loyal to his family and did not ask much for himself in return. She encouraged his determination to excel in school and whatever else he set himself to do and also appreciated — and came to depend on — the money he contributed to the family. He was her 'wee man' and as such, deserved special treatment by the others, although this did not always sit well with Jim's two eldest sisters, Meg and Cis. Doll never complained about the way her mother did things.

One Saturday as Jim returned from his grocery deliveries with Mary sucking on a toffee beside him, he opened the door to be greeted by Bill and Jock racing toward him.

"Do you have treats today?"

"Jock, don't be so rough on Bill," Cissy hollered.

"He pushed me," Bill began to wail.

"Fix your brother a jelly piece," Martha called to Meg. "The lad is probably starving."

"I've been at the mill for over a year now," protested Meg. "I give you part of my pay. I don't spend much on myself, except for going out to the dances with my friends. I help with all of the household chores and you want me to cater to him."

"*Haud yer wheesht!*" her mother countered. "He's a lad and you're just a lass and you'll do what I say."

"It's not fair. I'm the eldest in the family, the first to get . . ."

"It has nothing to do with fairness. It is just the way it is and you need to remember that," Martha interjected, raising her voice to make certain she was being heard over the din in the corner where Cissy was chastising Jock and Bill, whose predictable wailing reverberated off the stone walls. After a quick glance toward the source of the racket, Martha returned her stern eyes to Meg.

"Cissy," she ordered, "wipe Bill's nose and take the weans outside." She continued with Meg. "As long as you're under this roof, you'll do as you're told, lassie."

"Well, I hope it won't be long. This place is too crowded and you keep having more babies."

"Don't get fresh with me. Make the jelly piece for your brother," Martha snapped back.

"I don't want to be stuck here the rest of my life. I want to see something more," Meg mumbled as she turned to the table.

"You're already seeing too much. The way you're acting with your dancing and parties, I wonder if you've learned anything about proper living."

"I'm just having fun and not doing anything wrong, Mother," Meg protested while she got the knife out and vigorously attacked the bread.

"You'll come to no good if you keep this up," sighed Martha with some exasperation. "I need you to be here and help with the weans. Sometimes I think you just let the devil take over . . . It would do you good to start back to church and learn more about what your sinful living will lead to."

Meg ignored her mother and cut a jagged slab from the half-loaf on the table. She dipped a spoon into the stone jar of Robertson's Golden Shred marmalade and smeared a small amount on the slice. As she plunked it on the table in front of her grinning young brother, she wondered what her mother might do if she knew her special 'wee man' was smoking. She has blinders on when it comes to that boy. Anyone with a nose can smell the smoke on him.

Martha shrugged, turned and walked slowly to the stove, not wanting her daughter to see the moisture forming in her eyes. She thoroughly stirred the pot of simmering soup.

It upset Martha, a strong Bible-believing Baptist, to think that her daughter went to dances and parties, where she might be tempted to drink and participate in God knows what other sinful activities. In spite of Jimmy's backsliding ways, she made sure her children attended church each week and followed the strict rules laid out for clean and righteous living. The minister's sermons always seemed to reinforce her belief that you had to be a believer to get to Heaven, as she was determined that her family would. As the children grew older, they learned that Sunday was filled with 'no's: no whistling, no playing cards, no using scissors, no singing worldly tunes, no books to be read except for the Bible and no unnecessary chores to be done. They had to attend the Sabbath school and, as soon as they were old enough, joined the Band of Hope in order to get an early temperance education. Often the children would be singing the band theme song while their father dallied at the pub.

> "I promise here by Grace divine to drink no spirits, ale, or wine
> Nor or will I buy or sell or give strong drink to others while I live.
> For my own good this Pledge I take but also for my neighbour's sake
> And this my strong resolve shall be: No drink, no drink, no drink for me."

Jimmy, with staid Presbyterian roots, did not have the same passion for the church. After the family moved to Tillicoultry, he stayed home on Sundays with young Jim while Martha went off to church with the three girls, leaving Jim with the reminder of what a poor example he was for the children.

"They'll soon learn what a heathen you are."

Having waited eight years for the arrival of his firstborn son, Jimmy was determined to spend as much time with him as he could and cherished those Sunday mornings when the two could be together until the lad was old enough to join the trek to church. The Sunday pattern continued as new babies came, and as soon as each reached the age of three, they joined the rest in the Sunday ritual.

As Jim grew older, Jimmy looked for new ways to spend time with him to share stories and wisdom. While he knew that some of his choices and his lack of ambition had hindered him from moving ahead, he was determined that his son would have the opportunity to do better, but it would not be in the mines. When the boy was old enough, father and son began to spend Sunday afternoons fishing together on the banks of the Devon.

Martha had reluctantly relented in her adamant stance about appropriate Sunday activity as long as Jim went to church with her in the morning. While she never gave overt permission for their outings, her tacit silence let everyone know that it was acceptable, especially if it meant a tastier and more wholesome supper. Besides, the good Lord had a special place for fishermen.

Jimmy was patient with his son as he coached him in the fine art of casting, week by week.

"Bring your arm a little farther back, laddie. Gently swing it around like this," he offered, demonstrating the smooth action of his arm by drawing it back and then gently forward over his head.

Jim emulated his father as best he could and, while at first he could not match his father's rhythm or achieve the same distance in his cast, he continued to improve.

"Well done," said Jimmy, smiling broadly at his son when the cast landed mid-stream.

He noticed that one of Jim's ears was very red, but said nothing.

"Sometimes I practice with a willow when I'm walking home," Jim replied.

"Good. Let's see if we can catch a bit of supper," Jimmy said, "It will make your mother happy. What's the matter with your lug? It looks like someone bashed you again."

Jim reached up to touch his tender ear.

"It's those big boys," he said. "The young ones pick a fight and then call in their older brothers to finish it. I'm no match for them. It's not easy with no big brothers. But I tell you, when Jock and Bill get to school, they'll have nothing to worry about."

"You've a brave heart and you can stand up to those bullies," Jimmy said encouragingly.

"How can I do that? They are bigger and stronger than I am."

"You've got a head and good boots. Use them."

Jim looked down at his tackedy boots.

"What do you mean?"

"Just what I said," answered his father. "Your head and your boots can be good weapons. Courage and cunning are more important than size."

He cast his line, letting the fly float far out on the surface of the water, retrieved it and started all over again. Then he turned back to his son and pointed across the river toward Stirling Castle in the distance.

"Do you remember what happened in that field about 400 years ago?"

"The battle of Bannockburn?" Jim replied, hoping he was right.

"Yes. One of the greatest fights in Scottish history. Remember the little verse I taught you about Bruce and de Bohun?"

Jim beamed and began to recite:

"Bruce and de Bohun were fighting for the croon
Bruce took his battle-axe and knocked de Bohun doon . . ."

"You've a quick mind, Hamish," said his father with a smile.

Jim liked it when his father called him 'Hamish'. It was as close as he ever got to showing any tenderness toward him. He also knew that the verse was a prelude to another history lesson.

"The evening before the battle of Bannockburn," Jimmy began, sure enough, "Bruce was riding his pony, not wearing armour but carrying his battle-axe, when Sir Henry de Bohun, an English knight in full armour, riding his huge war-horse, recognized the gold coronet on Bruce's headband and charged at full speed. Just as he closed in, Bruce turned his nimble pony aside, avoided the thrust of de Bohun's lance, stood up full height in his saddle and with one blow of his axe to de Bohun's helmet, felled him."

"He was a very brave man." Jim cheered.

"What Bruce did the next morning was even braver," Jimmy went on. "The English outnumbered the Scots three to one and so it all depended on leadership and determination, not on size."

He paused for a moment to let those words sink in. He continued to cast and reel in his line as he talked. Each part of the story was like a fresh cast followed by the slow reeling in.

"The next morning as the two armies prepared for battle, the Scots went down on their knees to pray, and King Edward said, 'See, they are kneeling to ask for pardon.' His companion replied, 'They are asking pardon from God, not from us.'"

"Why did they need pardon?" Jim asked. "Was God mad at them?"

"No, those heathen Sassenachs just didn't understand how devout we Scots are. Having faith in God is more than the fire and brimstone you hear at church with your mother."

"When I'm older, I don't think I will keep on going to the church," Jim said.

"Don't tell your mother that."

Jim nodded.

"What did they do after they prayed?" he asked.

"Well, the battle began and raged for many hours. It was not looking good for us until some of the local people and the soldiers' servants came to help, waving banners as they screamed down the hill. The English thought it was another army and our soldiers took advantage of their confusion and turned the tide of battle. King Edward and his troops fled toward . . ." He broke from his story. "I've got a strike! Fetch my creel, lad."

Jim watched as his father carefully played a beautiful brown trout to shore. Before Jimmy disengaged the hook, he invited his son to use the cudgel to club the fish. With the hook removed, Jimmy held up his prize.

"This is a fine fish, lad," he proudly announced. "And you'll not find a brown trout anywhere else in the world. Did you know that?"

"Yes, you told me that many times."

Plopping the dead fish into the creel, Jimmy carefully anchored the basket in the river to let the water flow through it and keep his catch fresh.

"So you see, laddie," Jimmy smiled, "size and strength don't always win the day."

Jim liked listening to his father, whether it was recounting some gem from history, reciting a Burns poem or singing one his favourite folk tunes. Often when he was alone, he would sing the songs or recite the poetry and think about how good life would be if he could combine hard work and leisure fishing.

Chapter Five
A Wedding in San Giovanni
June 1911

Angelina slowly scanned the kitchen as she stood at the stove, stirring the polenta pot. She had grown up in this place, learned to cook with her father, served meals to her family and now things were about to change. When her own apartment had been built, she continued to cook in the main kitchen, telling her father that it was larger, better equipped than hers and familiar. She had not wanted her father to get too used to having her live independently, knowing that one day she and Francesco would move on. She smiled down at three-year-old Maria as she piled wooden blocks into a tower in the corner.

Putting down the wooden spoon she had been using to stir the bubbling yellow mass, she unfolded the letter again, and for the fifth time she read the second paragraph. She had already committed it to memory after Luigi had dropped it off on his way back from the village.

I have finally saved enough to come home and marry you. I will be sailing from New York to Le Havre and take the train through the Brenner Pass again. I should be there around the middle of May.

He went on to say that he would be working for his father and that there was space for the three of them to live there.

She had been waiting for those words for three years and now it was happening. *I won't have to be doing this much longer.*

"Your papa is coming home," she said to Maria. "Won't it be wonderful for the three of us to finally be together in our own place?"

"But what about Nonno and my uncles?'

"They will stay here and you will still see them all the time."

She turned back to the stove, preparing supper the way she had done since her teen years.

When Valentino and the boys returned from the vineyard and were seated at the table, the letter was the centre of conversation.

"Well, what did the scoundrel say this time?" Luigi asked.

"Do you really want to know?" she snapped back.

"Angelina," her father said. "Why are you so abrupt with your brother? You know all of us care about what is happening with you."

"I know that, but it is my life and you all hover around me like I am helpless." She looked from one brother to the other. "It is time you started taking care of yourselves. Luigi, you are twenty-eight, and Benedetto, you are thirty. Most men are married and having their own family by now. I won't be around much longer for you to fuss over."

"What do you mean?" asked Benedetto.

"You want to know what is in my letter? Well *this* is what is in this letter."

The three men seemed glued to their chairs as Angelina read the pertinent paragraph to them.

"I will believe it when I see it," said Valentino. "Even if he does show up, what happens next?"

"We are going to get married right away. Francesco's mother has talked to Father Giuseppe and he will marry us in the church."

"The scoundrel's father probably made it worthwhile for the priest to forget about sin and please those who support him," said Benedetto.

"You were fine when Father Giuseppe baptized Maria. Help me with the polenta pot, Luigi," said Angelina. "Let's eat it while it is hot."

She did not want any further discussion about plans for the wedding.

• • •

Francesco, feeling downright miserable and sorry for himself, stood hunched over the ship's railing, folded arms cushioning his chest from the hard wood. The long arduous train trip back across the country had none of the adventure and had felt like ten times the discomfort of his journey to the new land three years earlier. This time the cars appeared noisier, draftier, the seats were certainly less comfortable and the people appeared less friendly. He had not enjoyed that part of his journey home. Now, after three days on board this rocking and rolling ship, he was spent. He had nothing left to heave. Only when he remained perfectly still and munched on the dry bread the steward had given him was he able to hold the nausea at bay and forestall another dash to the toilet.

Today the sun was shining brightly, yet the seas still carried the memory of an earlier storm and he was deep in lonely gloom. This trip was not at all like the one that had brought him to America three years earlier. Then he had been on an adventure and even the choppy March ocean had not had an adverse effect on him. Now, as he thought about what he would be facing on his return and tried to sort out just how he would deal with his father, Angelina's family and the priest, the *mal di mare* paralyzed him. He asked himself if it was really seasickness or dread that churned his stomach.

Once he was back on land and sitting in a comfortable rail car in the station at Le Havre, his stomach settled down. With a sumptuous meal in front of him, he looked around the dining car and thought about how well he had done and how good it was to be back in familiar surroundings. European trains were so much more civilized than those he had experienced in Canada, and he felt somewhat smug as he smiled at his reflection in the window. He was about to show everyone that, as a Linteris, a Friulan and a man of honour, he had returned to do the noble thing; claim his bride and take full responsibility as head of his own family.

When Francesco stepped down onto the platform at the Casarsa station, his siblings engulfed him, each wanting to be the first to touch him. His mother broke through the wall of children and threw her arms around her son.

Her lips directly to his ear, she said, "My boy, I am so happy to see you." Wiping tear-filled eyes, she turned to Umberto and urged, "He looks so good. Be kind to him."

Umberto joined the embrace and the three of them silently held each other. Then Umberto stepped back and took his son's hand.

"Welcome home," he said. "It has been too long."

Francesco did his best to stay focused on his family and return their greetings but his mind was on the one who was not there. Catherina, sensitive to her son's feelings, put her hand on Francesco's shoulder.

"She will see you tomorrow at the meeting about the wedding," she whispered. "Arturo went to see her father and he insists on having the council right away. You know there won't be any wedding unless Valentino agrees. It all has to be arranged for June."

Francesco said nothing as he patted his mother's hand. She knew him so well and understood what he was thinking. Custom required that the fathers of the bride and groom would meet with the groom in a men's only council to discuss details of the wedding and to offer formal agreement to the union. *A meeting tomorrow? Hardly off the train and already the traditions are taking over.*

"Can I drive?" Francesco asked.

Without waiting for a reply, he took the reins as his brother loaded his baggage into the family carriage and they all crammed into it for the two-mile ride home.

That evening centred on the distribution of the gifts Francesco had brought for each member of his family, his stories about Canada, and the tasting of the wine that had been produced in his absence. It was one of the most relaxing times the family had enjoyed in years.

Later, preparing for bed in the room he shared with Arturo, Francesco felt he could speak freely.

"Did you see Angelina?"

"She's a bit nervous about how things will go."

"What did she say about me?"

"She's excited but you know how shy she is."

"How about the baby?"

"She's quite the little girl. Came up to me and called me '*Barbe* Arturo'. I look forward to being her uncle."

"Did Valentino sound agreeable to the idea of the wedding?"

"I can't tell. At least he agreed to the meeting."

The next day the *concei di fame*, or family council, convened at Valentino's home. The host and his two sons, Luigi and Benedetto, sat on one side of the table facing Francesco, Umberto, Arturo and Ermete. When the wine glasses were filled with Valentino's homemade *refosco*, the men set to the task at hand.

Valentino began.

"You have asked to marry my daughter. Why should I agree to that?"

From there the debate went back and forth across the table, almost as though it were a trial with self-styled lawyers presenting arguments to support their case.

"He is a scoundrel who ran out on a pregnant woman."

"He is noble and fulfilling his duty."

"He wants to take her away from her family."

"He can provide a good home for her and their daughter."

"But he left her all alone and pregnant to live with the shame of having a baby with no husband."

"He did not abandon her. He resolved even more to do the right thing and return to his responsibilities," Umberto declared. "He has agreed to work for and live on our estate to make it easier for Angelina and the child."

"You can't take Maria away," said Luigi.

"You are most welcome to visit back and forth," said Umberto. "It is time our families were closer."

It was obvious that both families felt strongly about the future of the couple and their child and, as the wine warmed their stomachs, so the room temperature increased. When it appeared that everything that could be said had been said, Valentino rose nobly from his place at the head of the table. Very much aware that he was both judge and jury, he slowly and deliberately pronounced his verdict.

"It is out of love for my stubborn daughter and my precious granddaughter that I will reluctantly give my blessing to this marriage. May God forgive me if I am making a mistake."

Francesco slowly stood, walked hesitantly over to Valentino and offered his hand. Valentino looked intently into the young man's eyes, then, after a long pause, he reached out and took the extended hand.

"Thank you so much for your favourable decision," Francesco offered. "I will take care of and provide well for your daughter and your granddaughter." He turned to Arturo. "Please hand me the parcel." Taking it from his brother, Francesco, following tradition, presented a gift to his future father-in-law. "From my heart and my new country, I give you this."

Valentino accepted the offering and unwrapped a new felt hat. He forced a slight smile as he placed it on his head. The men around the table rose and applauded the agreement and the accompanying gift. Now the couple would be free to meet together and with the priest to discuss their future.

"Go and see them," Valentino offered with a wave toward the door. "They are waiting in their house."

With this part of the formalities over, Francesco relaxed a bit, grinned, nodded to his host, and quickly exited.

The door was open when he reached Angelina's. The two quickly embraced and would have held each other interminably had it not been for the tug on Francesco's pant leg and the well-rehearsed words, "Hello, Papa."

He looked at his daughter for the first time and could not say a word. He knelt and reached out to touch her head. Maria smiled.

"*We* are getting married," she said.

"Maria. My daughter," Francesco said softly and he took her in his arms.

His usually stoic demeanour softened, he held her quietly for a moment, fighting the tears, then reached into his jacket pocket.

"I have something for you."

"What is it?" the child asked impatiently.

He handed her a little package wrapped in soft tissue and tied beautifully with a yellow ribbon.

"Open it."

Francesco continued to hold her as she tore the package open.

"Oh look, Mama, earrings!" she exclaimed as she jumped down and ran to her mother.

"They're beautiful," said Angelina. "Let's put them in." As she replaced her daughter's earrings, she said tenderly, "Francesco, you remembered that I told you we had her ears pierced."

A few minutes later, with Maria off to show her prize to her *nonno* and the rest, Angelina suggested that she and Francesco go for a walk.

"A walk? I can think of better things to do."

He wrapped his arms around her and shuffled the two of them toward the other room.

"Francesco, we can't," she said, pulling back from his grasp. "If we are alone in here too long, Papa will find some excuse to send one of my brothers over."

She turned and headed for the front door.

"When are we going to spend time together? It's been over three years."

"After we're married."

"That's almost three weeks," he replied with exasperation.

He reluctantly followed her.

"Let's go," she said quickly, opening the outside door.

There was so much to talk about, considering the past three years and the wedding preparations. They paused at the edge of a meadow beyond the sight of the houses and Francesco took another package from his pocket.

"I brought you a gift for the wedding and I want you to see it."

Angelina carefully undid the ribbon, gently unfolded the tissue paper, opened the box inside and held up a ring.

"This is nice. What is it for?"

"It's your wedding ring!" he said proudly. "The latest . . ."

"That's all you think of me — to buy me a silver ring?"

"It's gold. White gold from America."

"I want a gold one!"

"That is very expensive gold!" he protested. "It's the latest thing!"

"Gold is gold and this is silver, I don't want it! The ring has to be gold!"

With that, she threw his precious ring over the fence and into the tall grass.

50   Belly of Blackness

Frank was dumbfounded as he looked over the fence and then back at his bride-to-be. How could he have been so stupid? Tradition said a gold ring, and to her this was not gold.

In frustration at himself more than her, he blurted out, "Angelina, you are a stubborn woman! I thought you would . . . but if it is yellow gold that you want, I will buy you a magnificent gold ring."

He searched in the grass for a few minutes and then, with Angelina expressing concern about their long absence, they reluctantly made their way home — without the ring.

Angelina and Francesco did not have much time alone together in the days leading up to the wedding. There was much to do and others were always around to assist them. Luigi and Benedetto were relentless in their self-appointed roles as sentries, forcing Francesco to keep a chaste curfew when making evening visits to their sister's home. He found time to buy a 'real' gold ring.

Francesco talked with his father about working on the estate until he had enough money to return to Canada, hopefully by the end of the year.

"It will be the same as before," Umberto said. "You will follow my rules and keep your mouth shut about politics." Francesco looked at his father and nodded.

Prior to Francesco's arrival, Umberto had asked Father Giuseppe to officiate at the wedding in the San Giovanni church on the morning of Sunday, June 18. He had agreed that he would do it at eleven, following the nine o'clock mass. He had also agreed that Angelina could wear her mother's wedding dress, including the veil, even though he did not agree with the superstition that the veil would shroud her from evil spirits. Later, when he had met with the couple to discuss details, he told them that they would have to make their confessions prior to the ceremony. Francesco had not attended church in his time away and was reluctant to get into the confessional and admit this to the priest — let alone the other obvious admissions that would be expected from him. He told Angelina he might just not attend.

"But you have to, Francesco," Angelina said.

"I don't see why anyone should have to tell their sins to another man."

"That's the way it is in the church, you know that."

"Yes, I know that, but what has that got to do with God?"

"He won't marry us if you don't."

"All right," he said at last. "But it will the last time I ever do it."

"Don't tell him that!" Angelina pleaded.

"I came all the way here to get married. I'm not going to mess it up."

Finally, the day of the wedding arrived. Ermete dropped the Linteris family off at the church and then headed out to the Clarots' to pick up Angelina and her family. Francesco had insisted that she arrive in the fine Linteris carriage drawn by the white stallion. While his parents mingled with the arriving guests, and his sisters flitted about in their finery, Francesco and Arturo walked around behind the church and through the grounds that bordered the cemetery.

"I haven't been here since Nonno's funeral," said Francesco.

"Why do you want to be here today?"

Francesco looked along at the wall containing the Linteris plaques and found that of his grandfather.

"He was a good man and understood me," he said.

"You should be thinking about life today, not death," Arturo offered.

"I am. He knew how to live and sometimes I get strength when I think of him."

"You're going to need it today," said Arturo, trying to break the mood that was becoming too serious for him. "Come on, we better get back and wait for Angelina."

As they walked, Arturo teased his brother with comments like, "Do you think she came to her senses?" and "I don't think she'll be here." Francesco hated all the traditions that surrounded this day, and while he knew that Angelina would be fashionably late, he was still upset as Arturo kept up the customary discouragements as they waited for the carriage.

The clopping of the horse announced the impending arrival of his bride. The two young men watched the dramatic approach.

"You," Arturo said, suddenly more serious, "are a very fortunate man."

After her father helped Angelina out of the carriage, Francesco handed her the bouquet of pink and red roses that symbolized the support of his family for the wedding.

Once the groom and his best man were in place, the music of the pipe organ announced the coming of the bride, and the congregation arose to greet her and her proud father. As Valentino walked his daughter down the aisle, he became aware of the disparity between the number of guests on his side of the church and that of the large Linteris family.

"There's a lot of them," he whispered into his daughter's ear.

Angelina turned slightly toward him, smiled and squeezed his arm.

The priest began to intone the wedding mass. Angelina relaxed, as the familiar Latin gave her a sense of comfort. Francesco, on the other hand, had not heard the language for over three years and had to stop his mind from taking him back to earlier days when he had endured the monotony of the mass. He focused on what was happening and was ready when the priest asked them if they had come of their own free will to the marriage. When they both replied in the affirmative, the priest went on to ask if they would honour and love one another as husband and wife and accept any children as a gift from God and bring them up in the Roman Catholic Church.

They both said "I do."

Following more prayers and a reading, the priest asked them to join their hands. He looked at Francesco.

"Do you," he asked, "take Angelina as your lawful wife, to have and to hold, from this day forward, for better or for worse, for richer or for poorer, in sickness and in health, to love and cherish until death do you part?"

"I do," replied Francesco firmly.

Then the priest repeated the question to Angelina.

She whispered, "I do."

Maria, who sat with her uncles Luigi and Benedetto during the ceremony, periodically whispered, "Are we married yet?" Luigi smiled at her each time and shook his head as he brought his forefinger to his mouth and whispered back, "Shhhh."

Finally, the priest offered a blessing on the couple and then blessed the proper yellow gold rings to emphasize that they were symbols of deep faith and peace. Following the benediction, the organ struck a chord and the couple followed the priest out into the brilliant sun of the noon hour. As the congregation fell in behind the newlyweds, Luigi swung Maria into the aisle.

"*Now*," he proclaimed, "you are married."

The bride and groom then led the gathered community to the village square and into the ballroom where the guests could greet the newlyweds and offer their congratulations. It was a good party, with abundant food and drink and dancing. Periodically someone would call out '*busa, busa*' and the bride and groom, pretending that it was a chore, would slowly and reluctantly stand up and kiss. Throughout the proceedings, guests would approach Angelina and slip money into the *borsa*, the small satin bag she carried for that purpose. The practical-minded Francesco embraced this custom, acknowledging the gifts with words and smiles of gratitude.

At the close of the reception, Angelina and Francesco handed each of the guests a bag of candied almonds, *bombon*, to represent the sweet and bitter aspects of life and a tangible reminder of the vows spoken earlier, "for better, for worse."

The newlyweds did not go on a honeymoon trip but instead moved into the rooms that had been prepared at his father's home and left little Maria with her nonno and uncles while they settled in. He had convinced Angelina that it would be better for them to live there because it was close to his work and far from the negative attitude her family had toward him. Valentino was not happy, knowing he would not see much of his daughter or granddaughter while they lived on the Linteris estate.

Francesco worked with his father and his brothers, taking his responsibilities seriously, holding his tongue to keep the peace and biding his time. Angelina helped her mother-in-law with the cooking and cleaning. Little Maria quickly adapted to the new living arrangements and was excited to have a *nonna*, another *nonno*, three new *agnes* and two more *barbes*. There was always someone to watch out for the little one, and her mother had a new freedom. Within a month of the wedding, Angelina was pregnant again and Francesco began talking about returning to Canada around the end of the year.

"I don't want to travel with two children," he reasoned. "I will work hard and we will save enough so that we can travel while you are able."

"It might be better to be here with our families. I know the midwife," Angelina countered.

Francesco would not hear of it and made the travel arrangements. He booked passage on *La Lorraine* which was departing from Le Havre on December 27, even though it would necessitate them leaving San Giovanni on the day after Christmas. The next few months sped by and soon the time of farewells arrived.

There was little joy at the train station. Beneath all the well-wishing was a feeling shared by both the Linteris and Clarot families — that this might be the last time they would see this little family.

The three of them stood at the rail as *La Lorraine* left the great port city of Le Havre and moved into the English Channel on its way to Ellis Island in New York. They were a handsome, well-dressed family, appearing far more affluent than the reality of their lives. Francesco had been diligent in saving his money while working at home in the period before and after the wedding. When he added the generous wedding gifts from his family and friends, he was able to purchase second-class tickets for the voyage, outfit his family in fine clothes from the best shops in Pordenone and still have a bit of a nest egg to set up house in Canada.

His tailored topcoat, worn over a three-piece suit, gave the impression that he was taller than his five-foot-five-inch height, and with his new fedora tilted at just the right angle to give the impression of self-assurance, he looked every bit the dashing businessman he hoped to portray. Angelina's long broadcloth coat with the beaver fur collar creatively hid the growing tummy. Her large plumed hat was fastened securely under her chin by a broad ribbon that delicately framed her beautiful round face. Three-year old Maria was wearing a woolen reefer coat over a serge sailor suit, a warm bonnet covering the ribbons her mother had carefully tied into her hair. She strutted around her parents, enjoying the sound her new button boots made on the deck.

The ship was barely underway when a strange ache began growing deep within Angelina. Her wistful gaze did not focus on the diminishing city or anything else within her view. Instead, she saw her father and her brothers eating in silence in the quiet of their childless home, none of them commenting on the tasteless meal Benedetto had half-heartedly thrown together — each thinking of the two who were missing. She could almost reach out and touch them as the vivid picture took shape in her mind. *Will I ever see you again?*

Maria was only interested in the ship and its passengers. With her back to the sea, she stared in awe at the massive red stacks towering above her.

"Mama, look at the smoke coming out of those big chimneys."

Her voice did not penetrate the world her mother was visiting.

"Mama, look! They're so big."

Francesco gazing vacantly at the vast ocean ahead suddenly snapped at his wife.

"Angelina, Maria is talking to you."

It was a long way back from San Giovanni and it took Angelina a couple of seconds to focus on her exuberant daughter.

"Yes, my darling?"

"The big chimneys!"

Following her daughter's gaze, Angelina drew Maria close and smiled down at her.

"Yes, I have never seen any chimneys that big."

She lowered her eyes to the sea.

Francesco, also gazing blankly at the sea, was thinking about what was waiting for them in America — a place of their own, an opportunity to get ahead and no wagging tongues to stir his blood — but he did not have the same enthusiasm that had buoyed him during his initial voyage, for he also knew that he was facing more backbreaking work in the blackness of the mines. When he had recounted his life in Canada to his family, he had glossed over what it was like being in the mine every day and gave no hint of the discrimination that was his lot. "Canada is a good land with many opportunities," had been the theme of his conversations. He would not lose face with them and let them know the truth and be the brunt of harsh reminders.

Angelina had not anticipated that the seasickness Francesco had alluded to would be so much worse than what she had experienced in the first three months of her pregnancy. Now in her sixth month, all she wanted to do was lie on the bed in the tiny cabin while Francesco, who would not succumb to the stormy sea, and Maria, whose excitement left no room for discom-fort, toured the ship and savoured the rich variety of foods offered in the dining room.

Most of the time, Angelina managed no more than dry bread, crackers and water. She just wanted to be left alone. In her gloomy despair, she wished that she was not on this ship, that she was not married and that she was not going to have another baby. The little one growing inside her also seemed to protest the journey with constant kicking and punching at her abdominal wall.

"Angelina, I have some simple food for you," Francesco offered on his way back from a sumptuous meal.

"I can't. I will only be sick again and it hurts so much," Angelina replied feebly.

"You have to feed the baby growing inside."

"I don't want to eat."

"Do you know how much the tickets cost for this cabin and all the meals? Then you say you don't feel well and can't eat."

"Francesco, it must be the baby. I am so sick."

"You are a weak woman. Come with me, Maria, we are going to eat."

Francesco and Maria mingled with other passengers and during the times she napped, he would tell Angelina what to expect when they got to America.

"One of the first things you will have to get used to is making all of our names English. I am called Frank and I think Angelina will be fine for you." He turned to Maria. "Maria, my precious, we are going to a new land and we will all have new names. Yours will be Mary. Can you say that?"

"Mary. Mary. Mary."

She began a little dance as she repeated her new name over and over.

"That sounds so strange, Francesco," Angelina offered.

She did not want to talk about America. She wanted to go home.

"That's the way it is."

She had been learning that it was of no use arguing with Francesco and, as she had no energy left anyway, she let it drop. *My little Maria. How harsh 'Mary' sounds. I will still call her Maria in my heart.*

The landing at Ellis Island was uneventful as Frank had already experienced the procedure and this time, with more English, he was able to navigate the bureaucratic immigration maze with ease, and had little difficulty with the train connections to Montreal. His previous experience on the Canadian train also made the long tedious journey a little easier for the family, as he was familiar with how to buy and cook food as they crossed the country. Angelina, whose longest train ride had been the trip to Le Havre, looked forlornly out of the window at the rocks and trees of the Canadian Shield.

"Is there no end to this country? We passed these same rocks and trees and lakes yesterday."

"We are almost halfway there. Now that you are feeling better, it will be good for you to get off and get some air every time the train stops. You can watch out for Maria while I buy the food. She wants everything in the shops when she comes with me."

"I am so stiff from the sitting."

"It will be good for you," he said curtly, and she acquiesced.

When they arrived in Coleman, they stayed with Emilio Giambri and his wife for a few days. Angelina relaxed in the presence of Frank's friends, who willingly helped with the care of Mary during the stay. Nevertheless, life in a coal-mining town was strange for Angelina; everything was dirty, none of the houses had paint and the cold world around her was gray and white. The constant smoke and ash from the coke ovens swirled over the town and settled on buildings, people and washing. She quickly realized that Frank did not like going to the mine each day. Even though he was one of the hardest workers, he was still treated as one of the lower classes and excluded from many interactions between the English-speaking workers.

He used the last of his savings to furnish the two-room cabin he rented for his family's first home on their own — the first step on his way up to the place where he would show them who Frank Linteris was.

However, Angelina was a shy woman with no understanding of English, and she was uncomfortable in situations where Frank was not present.

Her life centered on taking care of her small family's domestic needs and, because she had seldom experienced temperatures below freezing in all her twenty-seven years in Friulan, she was reluctant to leave the house in her first cold winter in Canada.

"Mary will freeze if we go outside," she said one day as she sat in front of the open oven door in her kitchen. Though she was still only speaking Furlan, she had made the adjustment to her daughter's new name.

"She'll get used to it," Frank said crisply. "That's why I ordered the long underwear for us all from Eaton's catalogue. Mary needs fresh air and so do you."

"It's all so strange here. I don't understand anyone."

"I will do the talking."

After that, whenever they were out he spoke for both of them and Angelina kept her place in the background. He handled the money, did the shopping and asserted his will as head of the household.

Once in a while when Frank was at work, Mrs. Panatti dropped by to see how Angelina was doing. Although Angelina had studied Italian in school and there are many similarities between it and Furlan, she still had difficulty understanding everything. However, she was grateful for the support, especially when the due date of her baby coincided with the first signs of spring. Mrs. Panatti, who had accompanied her on a visit to the doctor, offered to be with her during labour and to look after Mary on the days Frank worked while Angelina was in hospital.

When the time for the birth arrived, the labour was mercifully short and the baby arrived without too much pain.

"You have another daughter," the doctor announced.

"I want to call her Stella," Angelina told Frank when he visited that April evening.

"Yes, a good name," Frank replied.

"She will be my shining star."

Within a few weeks, it was obvious that there was something seriously wrong with Stella. She had trouble breathing and did not gain much weight. The doctor did what he could and finally suggested that, because he didn't have the proper equipment to test and care for the little girl in Coleman, the parents might think about taking her to the general hospital in Calgary. Angelina was distraught and told Frank that they had to do something.

"We will move to Calgary," he replied.

"What about your job?"

"We can sell our furniture and I will find work. It will be better for me if I can find a job above ground. Besides, our baby needs more than they have here."

Having made the decision, Frank quickly sold off their furnishings and they boarded the train for Calgary, where they rented a room in a downtown hotel. The doctors in Calgary discovered a hole in the baby's heart and informed the parents that, if she was fortunate, it might close by itself, but otherwise she would continue to decline. There was not much that could be done except to make the baby comfortable and wait. Angelina was heartbroken and inconsolable when Frank translated what the doctor was saying.

"I am not leaving my baby here to die."

"Angelina," he said tenderly, "we can see her every day."

"I'm taking my baby home and you can tell them that!" she said emphatically.

Frank sensed that this was one argument he could not win and accepted Angelina's decision. Frank informed the doctor, who agreed that it might be best for everyone if Stella was with her mother, and assured them that he would drop by regularly to check on the baby.

Frank was able to find a few odd jobs during this time, and avoided any talk about Stella's condition. As the weeks became months, Angelina felt more and more lonely as she sat with her baby. She had no one to talk to and letters from her father were not very supportive, since she did not tell him how Francesco was responding to the situation.

The hole did not heal and before Stella reached her sixth month birthday in October, she died quietly in the arms of her weeping mother.

"Why did God do this to us?" Angelina wailed.

"What's done is done. We have to get on with our lives."

Frank showed no understanding of the grief Angelina was experiencing. Through all the turmoil, Angelina had not given much thought to the fact that she was pregnant again, three months now. All he knew was that he needed to get on with life and be working full-time again.

"We have to move, Angelina. I need the work."

"But our baby is buried here."

"You have to forget about her and look after yourself and the next baby."

I can't think of another one when Stella is so recently gone. She could not tell Frank her fears that God had taken the baby because she had sinned when she got pregnant with Maria and that he, Frank, had given up on the church. She wondered how much more punishment she might have to endure until her sins were atoned, and was afraid to find a priest to talk to about it.

In early January 1913, Frank had an announcement.

"I found a job in a new mine in a place called Lovett. Now we can leave Calgary and go to where the work is."

This news didn't mean anything to Angelina. She had been in Canada for less than a year and was expected to move again. Stuffing her grief into a place deep inside, she said nothing and began packing their belongings into a steamer trunk. They took the train north to Edmonton, and then caught the westbound to Edson where they changed to a spur line that formed a Y, which ran south and west to service a number of new mining camps. These two rail lines, which brought supplies in and took out the mined coal, were the only link with the outside world and became known as the Alberta Coal Branch.

The Coal Branch was in an area designated as a forest reserve where permanent settlement was forbidden, so the mining companies constructed and rented out the dwellings in each camp. Lovett, at the end of one line, was a typical company town with thirty identical new cabins strung out along both sides of the road leading from the railway to the mine. Twenty-six of these were rented to miners; the other four housed the hospital, a store and two offices for mine officials. The Mines Act required each colliery to have a first-aid hospital in each camp, and a doctor from Edson made regular trips to the camps.

Frank and his family were among the first to arrive in the new mining camp and, when he had gone to work for his first shift, Angelina looked out at the bleak winter landscape and yearned for her home in San Giovanni. With a deep sigh, she turned and began her household chores. In the afternoon as she began supper preparations, she discovered that she had forgotten to ask Frank to pick up tomato sauce when he had done the shopping on their arrival.

"Maria, come here please," Angelina called to her four-year-old daughter.

"I'm Mary. Papa says so."

"What Papa says . . ."

Angelina dared not say any more about what she thought about the names. It would only make Francesco more irritated if Maria were to tell him. He wanted to be more like the Englishers and, with Angelina stubbornly refusing to learn the language, he continued to be the public face for the family. Whenever she was forced to go out, she had to rely on her daughter to translate her requests, as Mary had quickly picked up the language by playing with other children and having her father read to her. She knew all the words necessary to ask the grocer for what her mother needed.

"Mary, we have to go to the store."

"I'm playing. Can't you go . . . ?"

"No! Come here and get your coat on. You have to tell the man what I want."

"But I can tell you the . . ."

"I told you to be quiet," Angelina said firmly.

She began to help her daughter into her boots and coat.

Together the two of them set off on the errand, with Mary searching for virgin snow in which to make tracks and Angelina walking cautiously, her growing belly affecting her balance on the hard-packed snowy roadway. She was worried that there might be something wrong with this baby too. Fortunately, Maria's playful leaping and sliding in the snow took her attention away from her negative feelings for a bit and they trudged to the store and home in the brisk air.

Just as the first signs of spring — a muddy road and a dirty floor — promised better days ahead, the child in her womb announced its impending appearance. One of their neighbours had helped deliver a number of babies and assisted in the birth in the little cabin hospital in Lovett on April 20, 1913. The baby was a healthy-looking little boy, whom they named Peter. Frank was proud to have a son and invited a few of his coworkers over for food and wine. It was one of the first celebrations they had hosted since moving there, and he had prepared a huge meal that impressed the neighbours. He liked to cook and basked in the accolades that came his way.

In September, Frank announced that he could do better in another new mining town run by the Yellowhead Coal Company, back along the line between Cadomin and Robb. The Yellowhead settlement had forty identical cabins, one of which would be the new home of the Linteris family.

"Frank, I don't want to move again so soon," said Angelina.

"We are moving, as it will be a better job for me and I will make more money."

"But it is so hard to keep getting used to new places."

"Will you be quiet!" he hollered at her. "All you do is complain. We are moving."

Angelina reluctantly resigned herself to the pattern that was forming. When Frank made up his mind, it was final and there was no room for any questions. He had less patience than earlier in the marriage and she did not want to hear him hollering all the time. He did not understand how much she missed Stella and she knew that any mention of her brought on more harsh words. She reverted to the silent docile cook she had been in San Giovanni.

By the end of January 1914, Angelina knew that there was another piece to the pattern of their moves — she was pregnant again. This time she would have a baby in the fall. Mary, who was now five, was a big help in amusing her baby brother Pete while Angelina went about her chores. Frank continued saving money for the day when he might be able to purchase his own home and be more like the Englishers, and he continued to read western novels and practice his English with Mary and the men at work. Angelina refused to participate.

On September 1, 1914, Angelina's labour started before she could get to the little hospital in the Yellowhead camp. Frank rushed out and located the community nurse, who called the doctor in Edson, then relayed the information that the doctor would come directly to their home as soon as he arrived in town.

The doctor, an elderly man, travelled by rail on a speeder, which was simply a seat and a one-cylinder engine on an open platform with four steel wheels underneath to run along the tracks. A section worker controlled the speeder as it putted its way through the quiet countryside, while the doctor warmed himself with nips of whiskey on the chilly ride. When he arrived drunk in the Yellowhead camp, he was in time to witness the nurse, who had experience as a midwife, delivering a little girl. After cutting the cord and checking to see that the new mom and baby were fine, he waved a cheery goodbye and left the nurse to tend to the clean-up. The trip must have been even nippier on the way home, as he neglected to register the baby when he sobered up the next day. Angelina and Frank chose a popular Italian name, Ida, for their latest child.

The next summer, a few months after Angelina became pregnant again, Frank decided to move to Fernie, BC, just across the border from Coleman. Fernie was a much larger community than any of those in which they had lived for the past few years and the family quickly settled in. The town, destroyed by a devastating forest fire in 1908, had been completely rebuilt, with most of the downtown edifices constructed of fire-resistant materials such as brick and stone.

In her first winter there, Angelina had little time or energy for anything more than looking after children, cooking meals and cleaning house while her abdomen expanded. Finally, on January 8, 1916, she gave birth to another girl. More than three years had passed since the death of their second child and Angelina decided she wanted to name this little baby Stella — a new star in her life. Francesco did not contest the suggestion, hoping this would finally put an end to Angelina's references to their dead daughter. He was preoccupied with the news of what was happening back home. The area around San Giovanni was a major battleground and he worried about the fate of his family. Angelina could not understand the words Francesco heard on the radio or read in the newspaper but she knew they made him angry. His anger was not only directed toward the situation in Friulan, but also at his wife who, because of her lack of English, could not discuss events with him.

What kind of stupid woman did I marry? he wondered.

Chapter Six
In His Father's Footsteps

The Elliot family continued to increase in numbers and, other than the three-year gap between the birth of Bill in 1914 and Robert in 1917, the pattern of biennial births continued with the birth of David in 1919. During this period, the older girls — because they were girls — were expected to help their mother with the household chores and the care of the younger children. Cis and Doll did not complain about this, but when Meg began work in the knitting mill and agreed to give her mother part of her salary, she balked at being expected to continue with a full load of responsibilities at home.

"I work hard all day and would like to relax with my friends after work," she protested when her mother told her that she was not pulling her weight.

"We all work hard, lassie," he mother replied. "And we would all like to have it easier, but that is not the way life is."

"Well, some of the girls at work are going to the show at the Sauchie hall on Saturday, and I am going with them."

"I don't want you hanging around that den of iniquity."

"For God's sake, Mother, it's the Social Club, all the young people go there."

"Don't use the Lord's name like that. Och, why can't you be like your sisters?"

"These are not my children and I am not going to spend my life looking after yours."

Meg did go to Sauchie that Saturday and the next and the next after that.

"What do you do every Saturday?" Cis asked one Sunday morning as the girls were cleaning up after breakfast.

"There is this troupe of actors from Glasgow, who put on shows, and then we dance afterward. Oh, Cis, it is so much fun. And there is one actor that is very special."

"What do you mean?"

"Well, the last couple of times, he has asked me to dance. And we've shared a lunch together when they take a break. Don't tell Mother."

Martha was not happy about Meg's choice for Saturday nights.

"You are tempting Satan with your behaviour, lass. No good will come out of you hanging about with that lot in Sauchie."

"I know what I am doing, mother. We are all just having fun."

One Monday evening in the spring of 1921, as she was helping to clean up the results of another chaotic supper, Meg, now twenty, told her family she had an announcement.

"I have quit my job at the mill and I'm moving to Glasgow."

"What are you getting at, lass?" her mother asked. "You can't just up and go like that."

"I've been thinking about it for a while now, and it's what I want to do."

"Does it have anything to do with that actor . . . oops," blurted Meg, covering her mouth with her hands when she realized what she'd let slip.

"What's this about an actor?" her mother said.

Meg glared at her younger sister for a moment, then slowly turned to face her mother.

"His name is Henry Cushley and he is very nice. He told me he could help me find work in Glasgow."

"I don't believe this," Martha replied. "You're going off with a stranger? What a scandal!"

"Och, Mother, it's not like that. One of the members of the troupe offered me a room in her parents' home."

"But we need you here. We've all these mouths to feed."

"Cis is working now and Doll will start this year. I have made up my mind."

"Jimmy!" Martha yelled at her husband. "Do something about this girl."

"It is probably time for her to be out on her own. We can ask Liz to keep an eye on her."

"You know what your family thinks about us. We are dirt to them."

"Liz is different," Jimmy continued. "Meg, you make sure you look your aunt up."

The parents' response covered the deep grief welling up inside them.

She might be lost if she went to the city, but she would do what she would do. So Meg left, determined to make a good life for herself.

Jim finished the highest level of public school that was available to him in June 1921 and his teacher recommended him for advancement. On the last day of the school term, he stood in the front of his class at Miss Peebles request and had recited Burns' poem "A Man's A Man For a' That":

> *Is there for honest poverty*
> *That hings his head, an a' that?*
> *The coward slave, we pass him by -*
> *We dare be poor for a' that!*
> *For a' that, an a' that,*
> *Our toils obscure, an a' that,*
> *The rank is but the guinea's stamp,*
> *The man's the gowd for a' that.*
> *What though on hamely fare we dine,*
> *Wear hoddin grey, an a' that?*
> *Gie fools their silks, and knaves their wine -*
> *A man's a man for a' that.*

It was more than a recitation that day; it was almost a credo for this young man. He knew about "honest poverty," "toils obscure" and what it was to dine "on hamely fare". Robbie Burns had become one of his heroes and, coupled with his father's teachings, Jim had fixed firmly in his mind that what really matters is not what you have but who you are and how you live. He cherished the leather-bound edition of Burns' poetry given to him by his Aunt Liz on his twelfth birthday and often read it, memorizing his favourites like "Holy Willie's Prayer" and "To a Mouse".

If you had looked closely that day as he recited those famous lines, you would have seen the moistness in the eyes of the stern Miss Peebles. Another of her best students was leaving school and would not pursue further education. He had written a composition about his dream to become a doctor, but she knew there was no way his parents could afford to send him on to the next level of schooling. She had encouraged him to seek financial aid from the council and had given him a form to take home for his parents to fill out.

When Jim broached the subject at home, it added more fuel to the flames of the family battle that had begun following a recent letter from

Jimmy's sister, Liz. She had offered to have Jim come and live with her and John so that the boy might continue his education and follow his dream. Jim listened in silence as his parents went at it.

"We've one awa' now in the city, one getting married," said Martha. "I'll no have another leaving home. And him so young."

"But it'll be a chance for the lad to better himself," Jimmy offered calmly, trying to appease his wife. "I wish I had taken more schooling."

"Better himself?" Martha responded. "He would get too many uppity ideas from them. I'll no have *them* lording it over us."

"What about the council money?" Jim interjected. "Miss Peebles said that with my marks, I should get a scholarship."

"We're not going to the council with our hand out again," Martha said sternly. "I need you at home."

"I'll work steadier . . ." Jimmy began.

"I've heard that for twenty years. Do you think I'm daft or something? You've never amounted to much."

Jimmy knew that this would only get worse.

"You're right as usual, Mattie. But it's hard to admit that we can't manage."

Jim pushed back the urge to blurt out his protests as he eyed the unread application form on the table. He understood his mother's need to have him close by for financial support. He knew the depth of his parents' pride and how they would not want to go through the rigorous and humiliating means test that saw more applications rejected than accepted. It was demeaning enough to fill out forms for assistance from the council without adding the lengthy process set up by the school system for aid. Jim did not let his mother and father see his deep disappointment. For as far back as he could remember, none of his fantasies had ever come close to reality. Why should it be any different now? *But someday, I will be away from all of this.*

Although Jim was six months short of his fourteenth birthday, the minimum age for leaving school, he had completed the required level and qualified for an exemption that would allow him to begin working. Moreover, he knew that it would be easier to get a job than to obtain financial aid from the bureaucracy.

"We'll find you a trade so you can do better than I have," said Jimmy as they sat at the table over a cup of tea.

"Most of my friends have got jobs in the mines," Jim offered. "I can make a lot . . ."

"You are not going into the mines," interjected Martha. "So get that thought out of your head."

"With Meg away and Cissy and Geordie talking about getting married," Jim said, "and all the weans that keep coming, I can . . ."

"Jim, lad," Jimmy offered. "Your mother is right."

"But you are . . ."

"Enough!" roared Martha, "You're not going into the mines!"

Jimmy, with the help of a friend, found his son an apprenticeship with a plumber in Tillicoultry. Each day for the next five weeks, the boy reluctantly set off across the Devon to learn a trade that was not of his choosing. From the start, he did not like being an apprentice. It was too rigid. For the past five years, he had done his odd jobs the way he wanted to and, to top it off, apprentices were poorly paid. He did not want to spend the rest of his life in the kind of job that would keep him trapped in the style of life he had known from birth.

From as far back as he could remember, he had watched his mother work hard to keep the family fed and clothed. He saw how she knit, sewed, cooked and washed clothes with much more skill than many women with two arms. He marvelled at her ability to wrap a wet sheet around her arm stump, then slowly twist it with her right hand until it was dry enough to go into the rinse water, and then do it all over again before it went on the line. This image helped shape his determination to do what he could to make life better for her and for himself.

Now when he met up with his pals, who all seemed to be working in the mines, he was extremely frustrated that he could not participate in all of their activities. They made more money in a day than he did in a week and they could buy fish and chips or go to the pictures whenever they wanted. He couldn't. As the acknowledged leader of the group, he felt his place might be in jeopardy if he was not able to keep up with the rest. Something had to change and he began concoct a plan.

Each morning, he bounded out of bed with renewed energy, hurriedly slurped his porridge, picked up his jelly piece, shoved it into his pocket and headed out the door with a cheery farewell. Martha was pleased at her son's brighter spirit, grateful that he was finally settling in at the plumbing job. She quietly hummed as she rinsed out the jug in preparation for the visit of the milk wagon, hoping it would arrive before the rest of the children insisted on their breakfast. Jim was the one bright spot in her dreary world.

However, for the past five days, Jim had exited the house, crossed the bridge over the Devon, and then instead of continuing into Tillicoultry, had turned left onto the well-worn winding footpath that led through the bush along the riverbank to the Devonside Colliery.

"You here again, laddie?" It was more of a statement than a question from the manager.

"Yes, and I will keep coming until I get a job," replied the determined boy.

"Och, you're persistent. I can't be wasting my time with you hounding me every day about work. What's your name again?"

"Jim Elliot," he replied quickly, relieved that had not been asked his age this time. He did not like being perceived as young just because he was not as tall as some of his friends. It was times like this that he remembered the story his father had told him about Robert the Bruce and how it illustrated size not being important.

"Who do you belong to?"

"Jimmy's my father," he answered. "He works in Alloa. I don't think you know him."

Jim hoped he was right.

"He's in one of our company's mines then. You'll know how hard the work is."

"I do. Many of my pals are in the pits," Jim continued, encouraged that this was a longer conversation than any of the previous days. "I'm stronger than most of them."

"Do you think you could pull a car full of coal?" the manager queried, not sure why he was continuing the conversation.

"I've lugged a grocery cart since I was eight. Just give me a chance," said Jim, trying not to sound too boastful.

"Well, if your back is as strong as your tongue . . ." He paused for one more scan of the fidgeting lad. ". . . you might do fine. With a three-month strike and many of the men not returning, we are short. I've a job in haulage. You can start on Monday."

"Thank you, sir."

"Now go over to the lamp shack," the manager continued, motioning Jim toward the door. "Talk to Davie Patterson, the pit boss. He'll tell you what you need to know."

As he watched the boy walk quickly away, he smiled a little, admiring his spunk.

Davie explained to the eager boy that the work in haulage meant pushing full tubs of coal out from the digging area to the main line, where they would be connected to other tubs and hoisted up the long slope by cable.

"It's important that you get the empties back so the loaders can keep up with the cutters," Davie added.

He explained in detail what Jim would have to do to prepare for his first day and then dismissed him.

"You'll get the hang of it quickly. Meet me here at seven Monday morning."

With a cheery farewell, Jim left the mine, skirted Tillicoultry and headed up Alva Glen to find a spot where he could be alone and not worry about meeting anyone he knew. He stopped at the secluded pond below the falls, picked up a handful of stones and continued to mull over the implications of his new job. Thoughts skipped across his mind like the *chuckie stanes* he was bouncing off the surface of the smooth water. *One-two-three-four-five. That was a good one. Oops, that one kerplunked.* Just like some of his thoughts. *There'll be hell to pay when I tell my mother.* He lay back on a large rock, ate his lunch and shortly after noon decided that it would be a good time to go home. *My father will still be there prior to his afternoon visit to the pub, Cis and Doll are at work and Mary, Jock and Bill at school.*

"What are you doing home?" his startled mother asked as Jim entered the front door.

"I've just come from the mine. I've a job."

"You've what?" Jimmy exclaimed, looking up from his paper. "What about the plumbing?"

"I quit that a week ago. I've been going to the mine every day . . ."

His father was dumbfounded. With rising anger, he stood up to face the boy.

"You quit your job! Are you daft or something?"

Before he could say another word, Martha jumped in.

"What are you talking about? You're not going into the pit."

"But I can't . . ."

"You're not going and that's final!" Jimmy roared at his son.

"I've taken the job and I'm going!" Jim shouted back at him.

Martha looked at her 'wee man'.

"It's a man's work and you're too small for it."

Unfamiliar tears began to pool in her eyes.

"Look what you're doing to your mother, you blathering fool," Jimmy went on. "We've set you up in a respectable trade and you want to throw it all away. You'll not go into the mine!"

"It's not a fit place for you," Martha added.

"It's good enough for my father and . . ."

"Leave your father out of this," Martha snapped, brushing at her eyes with her apron.

"Well, I can work harder than him."

"You're a cocky bugger," Jimmy interjected.

"Maybe I am. But I'm going to work in the mine. And you won't stop me."

"Do something, Jimmy!" Martha ordered.

"What can I do? He's learned his stubbornness well — from you."

"So, now it's all my fault."

Martha knew that Jimmy would not stand up to the boy and deep down she also knew that when her eldest son said he would do something, he was true to his word — and the extra money would be a great help.

"You're his faither, lay down the law."

"He'll no' listen to me. You know that."

The argument continued to whir like a Catherine wheel with the words sparking as it went round and round with plenty of noise and heat, until it finally fizzled when the exasperated couple realized that they could not budge their strong-willed son.

Jim had stood silently when his parents turned on each other. He knew the pattern and simply waited for the bickering to burn itself out.

"I don't like it, lad," Jimmy said. "But if you do this, there'll be no complaining."

"You're so young," Martha sighed.

"I can do it. I'll show you," Jim said eagerly.

Recalling his conversation with the pit boss, he then told them that he would need a pail for his lunch and a tin flask for his tea, some old pants, heavy boots and a bit of leather sewn on the front of his cap to hold the tally lamp.

Later, when Doll came home from work, Jim recounted the events to her.

"There was hell a-popping today," he summed up. "I guess I *am* a cocky bugger."

The Devon mine had been operating for over a century as part of the Alloa Coal Company, but the operators had kept pace with mine safety requirements, and during the recent general strike had maintained and even upgraded equipment. Jimmy, who worked in a mine run by the same company, tried to assure Martha that at least it would be safe for their son.

On Monday morning, Jim collected his required gear. With his family crowded in the doorway, he said his goodbyes, and headed off to meet his pal, Bob Crawford, who had begun work a month earlier. The two had been pals for a couple of years in school and got on well together so there was always more to say than time to say it. Chatting their way along the Devon, Bob gave him a heads up on few things, like saving some of his tea until the end of the shift to rinse the coal dust from his mouth. The time passed quickly and they soon arrived at the mine. They entered the lamp room where the attendant handed each of them a tally lamp, newly filled with smelly whale oil and a fresh wick, and an extra can of oil. To Jim, the four-inch lamp looked like a miniature teapot with a white wick sticking out of its long spout. He hung it on his cap with the two prongs fitting snugly into the loop his mother had so grudgingly sewed on. From there the two boys went to the board where they picked up their number tags. Davie Patterson told Jim to put his name on a chart next to the number corresponding to his tag.

"Tie this around your neck and make sure you hang it here when you finish your shift," said Davie.

"What's it for?" Jim asked.

"Identification," said Davie. "Since I know where everyone is working, if there is an accident, we can quickly check to see who's in the area and know what we're up against in our search. But dinna worry. It's safe. You'll be working with Dave Hunter. He's been around a long time and is responsible for digging the coal in one of the rooms. Stick by him and he'll show you what to do when you get below. He'll be in the last car."

He pointed to where the men were climbing into a chain of coal cars. When all were settled, Davie signalled the winch man to ease up on the cable brake and allow the little train to roll down the sloping tracks into the black mine. Because the main corridor had small electric light bulbs encased in wire cages strung sparsely along its length, the miners were not in complete darkness as the cars rumbled slowly down the ever-descending tunnel toward their destination.

There was little light in the area in which Jim was to work, so the first thing Hunter had him do was light his lamp. After a brief explanation of the responsibilities this new miner would have, Hunter was off to his coal-cutting machine.

"Have a good day, laddie."

Soon, Jim was alone, not feeling quite as cocky as he had in familiar territory. He hoped he would remember his instructions. The gentle breeze of the relentlessly pumped air washed cold over his skin, and even his own breathing echoed in his ears. Water dripped from the ceiling and timbers creaked, but most alarming of all was the sound of scurrying rats. The black silence of the mine left so much room for sounds to fill it.

He was more frightened than he could ever remember and grateful that the blackness hid his fear from anyone who might be working nearby. His lamp was not much help, as the light from the candle-like flame illuminated only a small area on the floor below him and created eerie shadows wherever he looked. The roaring rumble of a coal-cutting machine began in the distance, and the scraping of shovels and picks in the branch corridors distracted him from the unfamiliar and frightening assault on his ears. Then more new sounds emerged from the void — the groan of steel wheels on steel tracks, the crunch of studded boots on the hard surface, the hoarse whisper of stiff work clothes caked with grime calling out the cadence of each scratchy step. He could just make out the spectre of a flame bouncing toward him, moving up and down in time with the harsh grating. Then the first tub appeared.

"You're the new lad," said the man behind the tub. "What's your name?"

"Jim," he replied, peering into the face of one who looked no older than he did.

"Well, Jim, everything will be fine if you keep me and others in empty tubs."

"I'll do that," said Jim, hoping he would be able to keep up as another loaded tub began to emerge along the track.

He soon found that he could push one full tub at a time to the end of the haulage but bring back two empties on some trips and thus make a faster return. It took him awhile to get used to working with the lamp flame only about four inches above the front of his cap. Following the example of the other miners, he twisted the cap a little to the left to make better use of the flame and, as his eyes adjusted, he was able to see the tracks in front of him. He soon developed a rhythm and relaxed a little as he went about his work, quietly singing some of his favourite folk songs to drown out some of the noise and keep his mind occupied.

"Time to eat," said one of the tub pushers as he and others came out of their areas and began to walk toward the main tunnel. "Get your tin. We eat down at the crosscut."

Jim retrieved his pail, which he had wrapped in his jacket as an extra precaution against hungry rats, and joined the others under the dim light bulb that marked the branch tunnel. He looked around for his pal, Bob, but he was working with another group further into the mine. Davie Patterson approached and, after a couple of questions about his first morning on the job, left Jim to get acquainted with the others. The conversation between the miners centred on the recently-ended strike and how good it was to be back at work. Jim listened intently, content to be part of this group and learn as much as he could from their experiences.

Jim hungrily tucked into his bread and jam sandwiches and drank his cold tea, and it was soon gone. The others had similar fare, knowing from experience that spending your day working in a crouching position, you get heartburn if your stomach is too full, so a few small sandwiches would have to suffice. Nothing fried was ever eaten down below or just before going down. Jim had his first mine lunch and began the routine that would become part of his work life.

He was no longer frightened by the environment by the end of that first shift, but his body was tired, his muscles ached and his clothes were wet from a combination of sweat and the constantly dripping water.

As they rode up the grade toward the mine entrance, the men in the cars ahead of him, silhouetted by the light at the end of the tunnel, appeared to be headless. The bowed heads may have indicated a prayer of gratitude for another safe return to the surface, but mostly they were the result of working bent over under the low ceiling during the long workday. When they emerged from the tunnel, the brilliant light of the late afternoon sun forced them to squint and then sneeze. After a day in the mine, it felt good. Jim rubbed his tongue over his teeth, creating saliva to help him spit out the dust that had collected in his throat. He opened his tea jar and sipped, rinsing out his mouth with the dregs.

Bob caught up to Jim and his crew as they got to the lamp house. Jim stuck his number tag on the appropriate nail and handed in his lamp.

Since the mine didn't have a washhouse, Jim and the other miners walked home with their damp dust-laden clothes chafing their itchy, grimy skin.

"The sun feels good," Bob said.

"Aye," replied Jim, "that chill gets into the bones. I'm awfully hungry."

"Get your mother to make an extra piece," offered Bob. "I like a bite at the end of the shift before we get into the cars. Helps clean out the mouth and stop the growling."

"When I get my pay on Friday, I'm going for chips on the way home," said Jim.

His thoughts shifted to what he would do when he had money, to those things now denied him by his poverty. He would also buy a mouth organ, a Hohner, get new shoes, go to the pictures and eat 'an elephant's sufficiency' of chips, as his father used to say.

"I said, you'll have more ideas than pay," his pal, Bob offered for the second time, the first having gone off into deaf space.

"I'll manage well. I won't need much for myself. I'll give my pay to my mother."

Soon they reached the Devonside Bridge, where Jim would cross the river. They made plans to meet in the morning. Jim had but one thought as he walked through Devonside: supper.

That first day also began a series of rituals at home. His anxious mother met him on the doorstep and bombarded him with questions about how he had fared as she helped him out of his dirty shirt. Doll immediately began to rub his back down with a piece of sacking. It was believed by Scottish miners, that if you washed your back with water, it would weaken you.

"I'm starving, Mother," Jim said. "That piece and jam did not last long."

"You'll have your bath first." Martha announced. "I've a big pot of kale ready."

"Sit still," said Dolly as she washed her brother, keeping his back dry. "Well, hurry up. I need to eat."

Once in the kitchen, he quickly shed the rest of his damp clothes and sat his naked body into a small tub of warm water, putting his feet into a second. Cissy took the washcloth, wet it thoroughly, rubbed the Castile soap into it, and then began to scrub her young brother's feet. She was not as gentle as Doll, but Jim enjoyed the pampering, and the bath quickly become a pleasant routine for him no matter which of the two had the foot-washing task.

With so many people in the small house, there was no room for embarrassment or modesty on Jim's part. That was just the way it was. As he grew older, he would move behind the curtained alcove and do his own washing, as did his older sisters. But on this first day he was exhausted and sat hunched over in the tub in silence, relishing the warm water and the hands of his older sister removing the grime. Martha, meanwhile, had hung Jim's mine clothes over a string near the stove, hoping they would be dry by morning. She would wash them in the left-over bathwater at the end of his workweek.

Robert, four, hung over the tub and rubbed the soapy grime on his own face.

"I'm a miner, too," he said.

Two-year-old David was content just to splash.

"You two get out of there and get back to the table," Martha hollered. "You're getting water all over the floor! Do you want me to thump your lugs?"

She waved her wooden spoon in their direction and the two little ones quickly heeded the command, drying their hands and obediently returning to the table where the other three children sat silently watching the bathing, knowing when it was wise to stay out of their mother's way.

After the bath ended, Jim donned his only other pants and shirt while his mother stirred the pot of vegetable soup simmering slowly on the coal cooker. It was now her turn to care for her son with nourishment for his groaning stomach.

"Sit down, lad. You've had a hard day."

"Aye, I have that."

It was all he could manage before he plunged his spoon into the simmering soup.

Jim settled into a routine of working five days each week and then playing football and going to dances on the weekends. The more experienced miners had shown him how to use Copenhagen snuff, a ground-up moist tobacco, which he inserted between his bottom lip and his teeth. The snuff produced saliva, which washed the coal dust from the inside of his mouth, allowing him to spit out the dirty saliva and retain the snuff. As a long-time smoker, it also gave him the lift that he enjoyed from his cigarettes.

The mine followed an annual-pay-raise-on-your-birthday policy, so on his fourteenth birthday in 1922, Jim received his first raise. Even with that, he was not satisfied with a job that was at the bottom of the wage scale, so he began pestering Davie to let him work at the coal face. He felt he was stronger than the other young men filling the tubs, even if they were bigger than he was. Davie had no more resistance to Jim's pestering than the mine manager had when he hired the boy. Although the lad was young and small, Davie eventually gave in and also agreed to mentor him as he moved to the coal face.

Jim's job now was to load about a hundred-weight (112 lbs.) into each of the small cars and then push them out of the 'lie' or work area to where they would be taken away to the haulage. He was able to keep up and, during breaks, learned more about how the mine worked and what other possibilities there might be for him as he grew older. Davie saw potential in the hard-working lad and began to teach him some of the basics of mining. He quickly learned how to nail up the cloth brattice screens that directed the airflow through the various mine tunnels, and how to shore up timbers and crossbeams to keep a safe ceiling above the 'lie'. The more he learned, the more he wanted to know. The more he knew, the more he looked for opportunities for advancement and thus apply his knowledge for gain.

Chapter Seven
Discrimination, Frustration and Anger

The house in Fernie was crowded and there appeared to be little room for privacy, but they were only there a few more months before Angelina would once more declare to Frank that she was pregnant.

"I don't know why you can't be more careful, you stupid woman," Frank replied to the announcement.

Angelina said nothing, as Frank's temper seemed to be growing with the family. She put her energy into her children and tried not to do anything to upset her husband.

Johnny arrived on a bright sunny day, February 2nd, 1917, and Fernie experienced six more weeks of winter. The family settled into a fairly comfortable life over the next four years and, in spite of the hard work involved in caring for five children and a demanding husband, Angelina had started to sing again in the kitchen. She began to creep out of her shy shell and risked getting to know a few of her Italian neighbours. She arranged for the children's baptism in the new Holy Family Roman Catholic Church and began attending mass on a regular basis. Frank, who believed that the rearing of the children was Angelina's responsibility, went along with her wishes about the baptisms but did not participate in the life of the parish. Angelina developed a routine that was comfortable and manageable and did not require learning much English, since most of their social activities centered on a small group of Italians.

Frank had lost some of his idealism by this time and began to realize that the opportunity to make it big in Canada was more of a dream than a reality for anyone who wasn't an Englisher. At thirty-two, he did not want to spend the rest of his working life in the coal mines and diligently saved every dollar he could toward that day when he would be able to buy a little farm and settle down "in my own place."

In the fall of 1918, when he heard that the Brazeau mine in Nordegg, Alberta, paid miners exceptionally well, he decided to quit his job in Fernie and move there.

"Frank, what are you doing?" Angelina asked. "Our life is good here."

"This is our chance to do better. It is good farming country up there and we can buy a place someday."

"You're going away and leaving me with five children?"

"I will find a house and come back for you as soon as I can."

Frank was able to find a two-bedroom house within a month and the family moved again. Ida and Stella slept behind a green curtain that portioned off part of the living room and Mary slept in a bedroom with her two little brothers.

At fourteen, Mary did housework for another family after school, and the woman she worked for taught her to crochet, a skill she passed along to her sister, Ida, now five years old. One day while practicing this new skill, Ida heard the other children playing outdoors. She hurriedly stuffed the crochet hook into her pocket and ran out to join them in a game of catch. At one point, as she reached down to catch a low ball, she impaled her arm on the crochet hook. Angelina rushed outside in response to the anguished screams and taking one look, knew that she could not remove the barbed hook. She ran down the street to the hospital with her screeching daughter in tow. Fortunately, the doctor was able to extricate the offending item and Ida returned home with a large white bandage as a trophy to show for her adventure. Angelina severely scolded Mary for giving her younger sister the hook. Ida was never to crochet again.

After a year in Nordegg, Frank announced that he had decided to use their savings and buy a small farm about two miles from Rocky Mountain House, hoping that one day he might turn it into a profitable operation and get out of the mines. The purchase included a modest two-storey house, a barn, a root cellar and some livestock that would keep the family in eggs, milk and chickens. Once more, the family was physically isolated from others and had to rely on Frank's weekly trip home to bring in supplies by train from Nordegg, ninety miles away. He controlled the money and on weekends did any extra shopping in Rocky Mountain House. He would hitch his two horses to the wagon and load up the children for a ride to town. Ida often helped by going into the store with her father while he selected the bulk items he felt Angelina would need, like navy beans at five cents a pound, tea for 75 cents a pound and prunes for 50 cents a pound.

Frank built a beehive oven beside the house so that Angelina could bake bread all year round without causing too much heat indoors in the summer. She learned how to stoke it with wood, get it hot, place the pans inside and wait for the heated bricks to bake the bread. Owing to Angelina's Roman Catholic background, the family never ate meat on a Friday, so, if Pete did not catch fresh trout, then it would be dried fish from the grocers. Ida endured 'fish Fridays', but had no interest in learning how to cook fish and only ate it when there was no other option.

Mary, Pete and Ida had to walk over two miles to the school in Rocky Mountain House when it opened in September 1920. When Ida started school, she was left-handed, which was not acceptable to the teachers. In order to force her to use her right hand, every time she picked up a pencil with the 'wrong' hand, the teacher would rap her sharply on the knuckles with a pointer. Gritting her teeth and fighting back the tears, she soon mastered printing and then writing with her 'right' hand. She had a great deal of spunk and was determined to show them all how capable she was.

The long walk to school was extremely arduous in the winter. Girls wore tights under their dresses to keep themselves warm, since they were not allowed to wear slacks. On the coldest days, it took a long time for the tingling in their legs to abate as they crowded around the kitchen stove at the end of a school day. The hand-knit mittens always seemed to allow the snow to collect around the wrists, which soon became chapped and painful. It was a difficult period for the family, as all of the work fell to Angelina and the children while Frank was in Nordegg. On weekends, they had even more chores when their father was around barking out orders. The children learned to live *arrangiarsi* during that four-year stretch at Rocky.

One fall day, when the garden produce was being stored in the root cellar, the door was accidentally left open. No one noticed the cow squirm in through the narrow door and begin to devour the fresh turnips. Later in the day, a steady moan emanating from the root cellar beckoned the family. They found the bloated cow trying to exit rear first, but unable to do so because of her increased girth. Pete crawled under her and tried to push on her head but that did nothing except increase the moaning. She would not budge — could not — budge. With the five children buzzing about making suggestions, Angelina looked at the hopeless situation and began to cry.

"What are we going to do? We have to get her out."

"Don't worry Mama," Mary said confidently. "We'll figure out something."

"She might die in there," Angelina continued. "Your papa will be so mad."

"Maybe we can push her sides in," Mary suggested. "Mama, pull on her tail. The rest of you start pushing!"

And with Angelina, using every ounce of strength she could muster, pulling vigorously on the tail, Pete pushing as hard as he could on the head, and the other four children leaning into the sides of the cow, she slowly inched back and finally popped into the open air. The giggling relief was shortened when Angelina admonished her children.

"Don't say a word to Papa! Make sure you shut that door from now on."

Over the years, Frank had become an increasingly angry man as he struggled against discrimination at work, the long weekly commute and the increasing responsibilities of his growing family. He was unhappy with the way the British prime minister had gone back on his word when the Treaty of Versailles was developed — cheating Italy out of some of the promises made earlier. He felt betrayed as he read of how Mussolini had deserted his socialist ideals and was leading the fascists in their take-over of Italy. His anger grew as he read letters from his mother about how much worse life was with the fascists in control than it had been under the Austrians.

The only place he had any authority was at home and the tenderness and caring that he had shown his wife prior to their marriage continued to diminish. He tightly controlled all family decisions and his verbal abuse of Angelina increased.

"What is the matter with you, woman?" he hollered one Friday night when he came home and found that the chicken coop had not been cleaned.

"The children had other chores and school work and didn't have time . . ."

"Don't make excuses!" he bellowed as he slapped her across the side of the head. "All you have to do is look after this place while I work hard all week."

He huffed off to see what else had not been done.

Angelina was stunned for a moment and turned quickly to the stove in case one of the children came in — her face burning as doubt simmered in her head. She had grown up with three men who treated her gently and, while it was the practice back home for many men to beat their wives, Frank had never hit her before.

He was kind to me in Italy. When we first came to Canada, he was mostly gentle although he did holler at me once in a while. Now he is getting worse. She quietly stirred her sauce. *Maybe my father was right. He is a scoundrel.*

82   Belly of Blackness

Angelina had been in contact with her father over the years but he continued to refuse to mention Frank's name in his letters. He had warned her when she left home that if she married that "good for nothing" she would be on her own. She could not tell him what was happening now. Her brothers had moved to California, where they had both married but they were determined to refrain from contacting Angelina as long as she was with Frank.

What have I done? But I must take care of my family.

Frank did not apologize for his action, as he believed it was the right of a husband to keep his wife in line any way he chose to do so.

Finally, in 1923, Frank had to admit that farming was not going to work for him and he decided to move on. His impetuous nature once again controlled his decision-making and he sold the farm for a pittance in order to get out. This time they moved to Lethbridge, where he found a job at the Galt Mining Company's Number 3 mine on the northern edge of the city. He bought a small house in a remote field and, once again, the children had quite a distance to walk to school and Angelina was isolated from her neighbours. She felt awfully alone as there was no one to talk to but the children, who she couldn't tell that their father had begun to hit her when he was very angry. He chose his times when the children were not around and they had no suspicions about his behaviour as their mother continued to sing while she cooked.

Within the year, he bought an old house in a community known as River Bottom, on the banks of the Oldman River below the mine near the railway trestle. It was closer to the school and the mine but the children did not like living there. The settlement had been developed around the first drift mine in the area, which had closed many years earlier, and had become a kind of catch-all for those who did not have the resources to live 'up' in the city. The steel trestle dominating the sky above had been completed by the Canadian Pacific Railway in 1909 and spanned one mile across the steep banks at a maximum height of 314 feet. It was the longest and highest bridge of its kind in North America and quite a sight for little children looking up from their lowly home.

The closest school was St. Basil, which had been built in 1914, and staffed by nuns under the tight, domineering control of Mother Superior, the principal. It was a long, winding, tiring trek up the hill, past the brewery to the top of the valley for Frank to go to work and the children to school. None of them looked forward to it. Frank did not like being in the mines and was angry that his farming stint had failed, and the children dreaded one more day in the rigid Catholic school. Ida did not like the idea that the children had to stop playing and come to attention whenever a priest came through the schoolyard. She also had difficulty with the way Mother Superior dominated their lives. How much easier and somehow shorter the trip home was at the end of the day. Children bounding and "I'll race you to that tree," all the way home to a mother who had filled the tiny home with wonderful cooking smells.

"Mama, Mother Superior is so mean," said Ida one day, as she savoured a fresh bun and wiped warm dripping butter from her chin. "I don't like her!"

"Ida, don't talk with your mouth full and don't say things like that about the holy woman."

"She is not holy, she is rotten!"

"Ida, you can't say things like that about Mother Superior. God is listening to you."

"Well, I hope He hears me. You know what she did today?"

"Don't talk and chew at the same time."

A quick swallow and Ida continued, "She chased Grino all over the room waving her strap in the air."

Grino's father had died earlier that year and, within six months, his mother had married a man who was very mean to the young boy.

"Why was she after Grino?"

"Just 'cause he swore at her. He is so young, Mama, and he has such a rotten life. How come Mother Superior can't be kinder to him?"

"Children shouldn't swear."

"But, Mama, it was so scary. She chased him up the aisle, and he ran the other way and was so upset, he just kept swearing. When he tried to get out one door, she got in the way so he ran to the other and she got in the way again. She was hollering at him and he was swearing and we were all crying. It was awful, Mama. It's a good thing she got tired and he ran home."

"Well, he shouldn't swear at Mother Superior."

Grino was locked in his basement for three days and fed only bread and water. He did not return to school for three weeks.

Ida's friend, Vera Kane, invited her to come home with her after school one day to look at their new telephone. They did not use it. They just looked at it and marvelled at how fortunate her family was to own one. On later visits, the girls became braver and Ida had the opportunity to pick up the phone and listen to other people's conversations on the party line.

Relief from the nuns came in the summer of 1926, when Frank decided to move once more to the Coal Branch. He had a job waiting in Mountain Park, at the end of the southwest branch of the rail line leading in from Edson. The first part of the trip was not too uncomfortable, but the final leg from Edson to Mountain Park took a day and a night in the old wooden coach with its slat seats. It was a relief when the slow-moving train finally reached its destination.

"I could walk faster than this," Pete said as the train crawled up the steep grade.

"You probably could," replied his father. "There are not many places in the world with tracks as steep as this."

Mountain Park, built on a dome-like rise, was the highest village in Canada at 6,200 feet above sea level, and the rail line servicing the mine was the highest in the country. It was another company town with a unique pattern of houses built on four parallel streets and along the road to the mine. Every second house had an upper storey, while the others were single-storey buildings, all of them painted white with green trim that gave the village a clean, uniform look — so different from the hodgepodge of many mining towns. The second floor of the two-storey houses was one long room — the length of the building — and was meant to provide living space for the single miners who made arrangements for board at some of the homes where the women liked to cook. The company also built a store, hospital, school and library, with the paint matching that of the residences. The houses, having been built on the mountainous rock, had no basements and could not even have outdoor toilets. The only exceptions were those built in the valley closer to the mine and occupied by the management families.

Because there was no way to dig toilet holes on the mountainside, the company had developed a system that involved using the dynamite boxes from the mine. The outhouses were all constructed with the same measurements and included a flap door at the back just large enough to accommodate an empty dynamite box. These were inserted like a drawer under the seat and then the flap was lowered. Every week or so around midnight the 'honey men', as they were called, drove their horse-drawn honey wagon from house to house and replaced the full boxes with empties. They then took their load out into a field about a mile from town, emptied the boxes, poured powdered lime into them and took them back to replace others that were full. No one really got to know the honey men, but everyone in town appreciated them and the fine job they did.

The valley houses had plumbing, while the rest relied on two 'water men' who filled the huge tank on their wagon at the river, and made regular deliveries to the residents. In the summer they used two horses, but in the winter the mountainous terrain meant it took four to pull their heavy load up the steep grade. Village children took great delight in sneaking up behind the lumbering wagon as it slowly made its way up from the river, and quietly undoing the plug on the water tank. The water froze along the roadway, creating a great run for sleighs. The watermen did not find any amusement in this activity, as they had to return and refill the wagons and then break in a new route up the side of the now-icy road. Servicing the upper floors was quite a chore as one of the men had to lug a huge hose up the stairs and hold the end in the storage barrel while his partner operated a hand pump on the wagon.

The family took some time in adjusting to the altitude, as they had to walk up and down the steep grades to get from their home to school or the store or down to the mine, which was located below the village on the banks of the McLeod River.

One day, when Ida was in grade four, she replied to a question from the teacher with "This here book is . . ."

"No, that's wrong, begin again."

"This here . . ."

"Ida, that's wrong, start again."

Ida had no idea what was wrong with what she was saying, so once more she picked up the book she had been asked to describe and said, "This here book is red and . . ."

"That's wrong!" screamed the teacher as she stormed over to Ida, grabbed her by the shoulders, and began to shake her.

Ida did not know what to say and the violent shaking continued until she threw up. After Ida had cleaned herself, the teacher carefully explained her grammar error. Ida did not tell her father or mother about the incident.

Ida began to develop an independent spirit and was determined to learn as much as she could in spite of the teacher. The next year, grade five, Ida began dreaming of becoming a teacher one day. She was one of the two top students in her class. When the teacher announced the top student each month, it would be Ida or Vera Scott, the daughter of the office manager at the mine. As the end of the year approached, the teacher, Miss Foster, announced to the class that Ida would be getting the award this year.

Ida flew home on wings of excitement. At last she would show them all how smart she was — even if her family was Italian.

But Vera Scott was heartbroken and went home in tears, upset that this lowly girl had come to town and taken away her status as top student.

The next day, the mine manager called Frank over at the end of his shift.

"Frank, I have just been talking to Mr. Scott and he tells me your daughter beat out Vera as top student this year."

"Yes, Ida is a smart girl."

"Well, Vera is very upset and I think she should get the award."

"But Miss Foster told Ida she was the best."

"Look, Frank, this is the way it is. If you want to keep working here, Vera has to be the best student this year. It would not look good if an Italian beat out a Scottish girl. What's it going to be?"

Once again, Frank had to swallow his pride and suppress his public anger as he agreed that the mine manager was right.

The next day in school Miss Foster tried to console Ida, but her "I am sorry, there is nothing I can do" only helped convince this eleven-year-old girl that she really didn't count for very much. This only deepened her resolve to show the world that she was as good as anyone else. Part of her rebellion was reflected in her decision to refer to her teacher as Miss Frosty from then on.

When the Italian kids got teased, Ida would fight back, not letting anyone know how ashamed she was to be Italian. Her one consolation was that the Slavic children were treated even more badly. Angelina, now a quiet, subdued person, still had not mastered English and asked her children to speak Italian in all conversations at home, but with so much discrimination toward Italian immigrants, the children were ashamed of their roots — wanted to deny them — and refused to talk Italian, speaking only English at home. It was so important for them to speak good English at school and with their friends.

In the summer of 1927, Frank began working with a new partner, a shy young Italian named Carl Franceschi. It only took two Sunday dinners with Carl in attendance for Mary to determine that this was the man for her.

"He is so polite and gentle, not like Papa," she said to Ida.

"But he is Italian."

"I don't care, I like him and I think he likes me."

And he did. During the summer and fall, the two spent as much time alone as they could and decided to marry in December, when they knew that Father Louis would be making his pre-Christmas visit.

Frank was overjoyed, but Angelina was a little skeptical as she knew how strong-willed Mary was and wondered what it might be like for her to marry an Italian. Nevertheless, they arranged to have the wedding. Father Louis was delighted to perform the ceremony as he knew the family, often having a meal with them.

He had begun serving the Coal Branch in 1919, and travelled from Edson to the end of both branch lines on a regular basis. He was a familiar sight as he trudged along the railway tracks with his heavy pack on his back, stopping at each section house and mining town to baptize, marry and often bury people.

As he took his leave following a delicious wedding supper, he clasped the hands of the newlyweds and wished them well.

Chapter Eight
Farewell to Scotland

In 1922, young Jim, now fourteen, jumped at the opportunity to work for Sandy Malcolm, an old man who had a contract for digging and blasting out the disk holes that provided space under the coal face. This allowed the arm of the coal-cutting machine to enter and chew its way along the seam. Sandy was a tough, crusty man who earned good money for this dirty job and was in turn generous to his helper. Jim shovelled mud out of the way, moving it along the corridor, and then helped Sandy with the drilling. Using an old ratchet machine, they would bore eight or nine holes about eight feet deep into the base of the coalface, fill them with powder and blast them off. The powder smoke and dust made it hard to breathe and often made Jim gag. When drilling at the bottom of a seam, water would collect and they would have to kneel in it, sometimes in a 36-inch-high space, and laboriously crank the wheel on the ratchet, forcing the drill bit into the coal. Since they did this when the regular miners were out of the mine, Jim had to switch to the afternoon shift, which also meant working Sundays.

Sandy looked forward to Saturday nights. He loved his brown malt porter, and thus on Sundays, he added his own powerful gasses to the mix in a space that had little air to begin with. On those Sabbath mornings, he often felt the effects of his imbibing and, with his left-leaning political beliefs, would say to Jim, "Just you sit there, laddie, and sing the 'Red Flag' and I'll drill." Jim loved it, for he enjoyed singing and would be able to get by with less work. He sang the verses, loud enough for Sandy to hear, and then with even more zest belted out the chorus each time he came to it.

> *"Then raise the scarlet standard high.*
> *Within its shade we'll live and die,*
> *Though cowards flinch and traitors sneer,*
> *We'll keep the red flag flying here."*

With nobody else around on a Sunday, Sandy used extra powder in the holes which blew the water and muck farther out, so they had much less shovelling to do. Each Friday at the end of the shift, Sandy would say to Jim, "Now remember, laddie, don't bring a piece box on Sunday. I'll look after you." Sandy loved fried *clootie* dumplings and, in the face of tradition and the possibility of adding even *more* gas to the situation, brought ample for the two of them every week.

Jim's fishing mornings with his father switched to the odd Saturday when Jimmy wasn't working and Jim didn't have other activities like soccer. The two of them would pack a jelly piece and a can of tea and continue their tradition of casting flies and sharing stories. Jim began to develop a talent for storytelling, with an increasing number of tales that got richer and spicier as he added Sandy's stories to those of his father. Jimmy knew Sandy, having often shared a pint with him, and had mixed feelings about some of the things his son was learning. Jim assured his father that he was doing well, not only because he worked hard but also because he had been fortunate to be with experienced men who encouraged him and continually talked of new opportunities.

"Don't you get tired of the fish and chips, lad?" his father asked as they were talking about Jim's plan to meet his pals for his favourite food that evening after supper.

"I don't think I will ever get enough," replied Jim. "You know I like trout and even fried kippers on a Saturday morning. I like a good feed of fish."

"Aye," said Jimmy with a smile. "You like a good feed of anything." And once again without apology he repeated the oft-shared saw, "Starved pups make hungry dogs."

Jim had heard this many times and did not reply, knowing his father was right.

Jim worked with Sandy for two years, but after he turned seventeen and still only five feet two inches tall, he once again decided that he wanted more than he had. He went to Davie Patterson and asked to be put on the long wall. This would mean more money because of the harder work involved.

"No, laddie," he said. "You will have to wait until you're twenty."

Jim reluctantly resigned himself to the situation and continued to work with Sandy, sustained by the dream that he would not be there forever. Most evenings Jim was happy to be out of the noisy house.

During this time, the infrequent letters from Meg indicated that Henry Cushley was becoming more and more a regular in her life.

"I hope she knows what she is doing," Martha said to Jimmy, following Meg's latest letter. "She's talking about getting married."

"We've never even met the gink," Jimmy said. "Is she having the wedding here?"

"She didn't say. It sounds like she wants it very soon."

"I hope she's not knocked-up like you were before we . . . "

"*Haud yer wheesht,*" Martha interrupted. "The bairns will hear you."

It was as Jimmy had suggested. Meg did marry Henry in Glasgow and in less than a year their baby, John, was born. Her letters home were full of enthusiasm about her life as a wife and mother.

"Are we ever going to see our grandson?" her mother wrote.

The question was answered two weeks later when Meg arrived at the door with her infant child.

"What a bonnie bairn," Jimmy said, greeting Meg and her baby.

"Where's his father?" Martha asked, looking past her daughter.

"He has a bit of a problem right now," answered Meg.

Meg handed her baby to Doll, who had rushed to greet her older sister.

"Come inside," her father said. "What kind of problem?"

"It seems as though he has been married before."

"You married a divorced man?" Martha asked. "You deserve better than that."

"Not exactly. Uh . . . Well . . ."

"What is it, lass?" Jimmy interrupted.

"He never got a divorce from his first wife, and the police came, and they took him away. He's now in jail, and I am all alone. I need you to look after John while I go back and wait for Henry . . ."

"Are you daft, lass?" her mother said. "Listen to the nonsense. You'll do nothing of the kind."

"But I love him and he loves me, and when he gets out, he will sort it."

She endured the "I told you so," "You should have known better," "Have you no shame, lass?" and other comments until her parents settled down and asked what she was going to do. She told them she couldn't manage a baby and work and that she wanted to return to Glasgow and wait for Henry. From the start, Meg had trouble accepting her son when she discovered his cleft palette. Once more, she was silent through another persistent diatribe. She bravely accepted her lot and when they had wound down, she asked again if they would look after John. All eyes turned to the subject of the conversation. Doll was cradling him in her arms and commenting on how *braw* he was. The little baby's face melted the ice in Martha's eyes. She could sense Meg's difficulty in accepting that her son was not perfect.

"We'll look after him, lass," Martha said. "We'll treat him like one of our own. Dinna worry."

She looked over at Jimmy who just shook his head slowly, anticipating another reason for a pint. After all, it had been four years since their ninth baby had come into the home and there were plenty of older ones to help. Of course, Jim would be exempt from this woman's work.

"Och, it feels like the Lord is punishing me for what I've done," Martha told Jimmy after Meg had gone.

"None of this is your fault, Mattie, Meg is very strong-minded and now we have another bairn to care for."

Jim had grown quite fond of dancing — with two older sisters teaching him — and began attending the weekly dance in the Dollar Hall with Doll and Cis, who was dating Geordie Bryce, a champion accordion player. Geordie often joined the fiddlers who provided music for the evening. It was good fun for the siblings, as they learned new steps and dances.

"Where do you get all of your life?" asked Doll one evening as they took a breather from the hectic pace. "You work so hard during the week."

"This gives me life," said Jim. "I love the music. When I hear it I can't keep my feet still. Are you rested enough? Let's go."

Doll nodded her assent and off they went to join the reel forming in front of them.

Jim continued to relentlessly nag the pit boss until January of 1926, when Jim turned eighteen. Davie once more gave in to the persistent pleading and agreed to let the lad have a go at the long wall. After three years with Sandy, Jim was well familiar with the long wall, for that is where they had been drilling in preparation for the coal-cutting machine. Now he could make more money by working with the crew that stripped the coal from the new cut. The machine did an undercut along the face of the coal to a depth of about six feet, with a series of sprags and wooden stakes wedged in under the coal to keep it from falling until the cut was completed. Then the sprags were removed and the undercut coal was made to fall by setting off a few strategically-placed explosive charges. When all of this was done, the strippers came in and shovelled coal onto a portable conveyor belt that was moved along as the coal-cutting machine brought down each long strip from the coal face. Each of the dozen men on the crew had to load an assigned ten yards of coal during their seven-hour shift. It was tough, back-breaking work as they swung the shovels into the three-foot-high and four-foot-deep pile. Jim worked as hard as any man on the line, determined that he would show them that size is not as important as heart. The more seasoned miners soon gained respect for him and accepted him as one of their own.

Shortly after he started this new job, one of the owners from the Alloa Coal Company was touring the mine with the pit boss and spotted Jim in the line.

"You're just a laddie," he said, looking down at the shortest person in his view.

"No, I am not just a laddie. I am a man!" Jim replied.

"I know you are doing a man's work, but how old are you?"

"I'm twenty."

"You don't look that old," replied the owner.

"He's not twenty. He's eighteen," interjected Davie Patterson.

"Is that right?" the owner said, warmly looking down at the sweating young man. "Well, I'm safe in saying that you are the youngest stripper with the Alloa Coal Company."

"Thank you, sir," said a beaming Jim.

As the visitors continued their tour, he turned and, with renewed vigour, pushed his shovel into the relentless pile.

"I'm proud of that young man," Davie remarked to the owner. "He is one of my hardest workers."

Life was not all hard work for Jim. He also played hard, having become part of a group, while still in school, that shared similar interests in football, dancing and eating chips. Since most young men moved about in gangs, Jim and six of his pals formed one of their own. They were not aggressive and did not initiate any fights, but found strength and support in each and did not back down when confronted by other gangs.

"I had no one around when I was younger," said Jim to his pals. "Let anyone try and lay a hand on my younger brothers."

The other members of the gang acted like an extended family when it came to confronting bullies, and thus all of their younger siblings experienced a security that had not been there for the gang.

Some of the larger gangs in the area had as many as a dozen members, but the group of seven, whose motto was 'Touch one, touch seven' took no nonsense from others and, as a result, commanded respect. They had all come from Tillicoultry and formed a bond while in school, spending many fun-filled hours in pick-up soccer games, a sport that needed no special equipment and gave many opportunities to get fresh air and exercise. Jim had learned to run and manoeuvre the ball while wearing his heavy tackedy boots. After he began to work and had spending money, he was able to purchase lighter boots for football and found he could run faster and dribble the ball more accurately. He also developed comfort and skill in heading the ball — his early education from his father about using your head and your feet had a positive spin-off on the soccer field.

During this period, a gymnasium opened in Alloa offering a wide variety of activities and appropriate equipment like dumbbells, weights, wrestling mats and a boxing ring. Over time, the older guys had learned to keep clear of Jim, who utilized his speed and cunning against their size and age. This new gym would give him an opportunity to develop new fighting skills, so he and his pals joined the gym and most of them took up boxing. The excellent coaching soon had the young men ready to enter county competitions. In his first fight, Jim won the novice event in his weight category. Then as his techniques improved so did his record, which included a win in the second round of a district competition over Tommy Spears, who went on to become the lightweight champion of the British Empire.

Jim's successes in the ring gave a new venue for him to express his cockiness. However, he was never content with the status quo and always sought to better himself, whether in the mine or anything else he took on. As his boxing skills improved, he continued to enter competitions, usually winning his section easily. His last competition was against a lanky six-footer in the same weight class who towered over his five-foot-seven-and-one-half inch frame. It was a frustrating fight for Jim, as his quickness and agility did him no good against his long-armed opponent, who simply held him off with one hand and hit him at will with the other. Try as he might though, the lanky lad could not put Jim on the canvas, on his way to a unanimous win on points.

"He sure pounded the hell out of me," Jim said through puffy lips as his trainer sponged his battered face.

"You put up a good fight, lad. You've nothing to be ashamed of," the trainer offered reassuringly while carefully removing Jim's gloves.

Jim sat quietly through the rest of the clean-up. The beating he had just taken brought back feelings of earlier days when the older boys had used their size and strength to best him. He did not like losing.

Negotiating his way home along the footpath from Alloa, he was hindered by his lack of vision in his swollen right eye and stumbled a few times in the darkness. Everything hurt. It was late on that Saturday night when he finally arrived and quietly climbed into bed without waking the two younger brothers who shared it with him.

He could sleep late, as he now had Sundays off with his new job. This favoured eldest son was still getting special treatment and Martha made sure he had his breakfast in bed on Sunday mornings. When Doll came in with his eggs, black pudding, and the fresh rolls she had just picked up at the bakers, she took one look at him and gasped.

"Mother," she bellowed, "come and look at Jim's face!"

Martha did look. What she saw aroused in her a mixture of rage, empathy, and a mother's protective love. No tears, just an unquestionable declaration which brought it all together, loudly and emphatically.

"No more fighting! No more boxing! And I don't want any argument."

With those words, Jim's boxing career was over.

His life as a miner took a turn for the worse on May 1, 1926, when the mine owners locked out the workers for refusing to accept the terms of a new contract that offered lower wages and an extra hour of work each day.

Jim talked about it all with his pal, Bob Crawford, as they stood on the Devon bridge, rolling cigarettes, following an angry walk back from the closed mine.

"I've just got this job going well and making better money and now they want to drop our wages from six pounds, eight shillings to six pounds, one shilling," Jim said.

"Aye," agreed Bob with increasing agitation. "It's criminal the way that lot operate, claiming to be going broke."

"I hear the Trade Union Council is going to call a general strike if the owners don't change their minds," said Jim. He licked the paper and rolled his creation carefully so as not to lose any precious tobacco out of the ends.

"The owners have the government on their side," said Bob, pausing to strike a match and offer Jim a light. "They aren't likely to pass any bills to help us."

"Well, I hope it won't be too long. I need the money. But until they get it settled, I'm going to have a long lie-in every morning."

They continued their walk to Tillicoultry, where they hoped to find enough of the gang around to get a bit of football going.

The general strike was called off on May 12, as management penalized many members of the supporting unions. The locked-out miners, feeling betrayed by the TUC and left alone in their struggle, voted to continue the strike.

This was a very tough time for the Elliots, as Cis was now married and no longer contributing to the family income. Fortunately Doll and Mary had both begun working in the mill when they had finished school. With the kind of belt-tightening the family was used to, they all struggled through together. Jim began to wonder if there might be something better than depending on the mines for his livelihood.

That summer, Jim played for the Devon Valley Rangers Football Club — named after their favourite Glasgow team. Recognizing his skills and desire to win, the coaches soon had him playing centre forward and outside right. Both of these striker positions demanded excellent foot and head dexterity, as there were as many goals scored with headers as there were by kicks. Jim used to like to tease one of his pals who was not as skilled at heading the ball, "Wee Eck had one kick at the ball today. It was a header and he missed."

Most of the gang made the team and enjoyed playing friendly matches in villages within a fifteen-mile radius. They used the bus system for transport as it was faster and more efficient than the horse-drawn wagons that had been used when they were younger. On game day, they usually stuffed sandwiches into their pockets for lunch. After the game they celebrated — win or lose — with fish and chips or black pudding and chips or mealy pudding and chips. It was a full day's outing and often other family members would join the lads to add more volume to the cheering. Although Jim was no longer boxing, he still went to the gym with his pals and took advantage of other equipment available to members of the athletic club. It was a great place to spend the days during the summer of the lockout.

His evenings continued to be filled by the local dances which were held almost every night of the week since people did not have to get up early. Jim and Doll continued to learn as many dances as they could and soon became very proficient. It was not long before they were taking prizes in local competitions, especially with the "Dashing White Sergeant" and "Strip the Willow". He loved the energy of the reels and strathspeys and tired out more than one partner in an evening. Dancing was so much more fun than boxing and carried none of the hazards. Jim could outlast most folk on the dance floor because of his stamina, built by hard work in the mines, activities in the gym and energetic playing on the soccer pitch. But in August when the strike was over, and he was back into the relentless ritual that was a miner's life, his dancing was restricted to Saturday nights once more.

During this period, Jim had started dating Mary Wilson, a young woman from Tillicoultry. It seemed so natural when he and Mary had begun to go out together; they had gone to the same school, her father was a miner and she loved to dance.

When Jim started work, he brought his paycheque home and handed it to his mother. She in turn gave him spending money and kept the rest for household support. That pattern continued as he got older and had better-paying jobs in the mines. Martha appreciated having Jim's money, especially after Cis and Geordie were married and had moved over to Alloa. Doll was generous but since she had begun dating John Cairns, everyone knew she would be the next to leave home. Mary, at sixteen, gave some of her meagre earnings to her mother, as did Jock, who had just started work following his fourteenth birthday. But there were still four other children to feed and clothe and their father was not very reliable even when the mine was working.

Jimmy did not like working in the mine and had little ambition to look for anything else, so his frequent "I'm no' well today" meant less money coming in and more going out as he looked for a cure in the pub.

Jim knew his mother saw him as the man of the house and, because she depended on him, he accepted his eldest son responsibilities without complaint. As long as he had enough to buy fish and chips, support his smoking habit, and, as he grew older, take a girl to a dance, he would do what had to be done. For the most part he was happy — well, almost happy. Often sitting at the table after supper with a cup of tea, he would look around at the crowded, noisy house that was his home and wonder about what life might be like for him now if . . . He would imagine life in Glasgow, going to school, studying hard to show the world that even a poor miner's son could become a well-respected doctor. *Then I could come back here and . . .* The bedlam around him usually cut short the daydreaming and he would return quietly to his reality.

His mother was always busy and never took a break from the constant cooking and cleaning and tending to children. The whole thing seemed like the waterwheel up Alva Glen, monotonously turning round and round but never going anywhere. As Jim watched her, he knew there had to be a better way to live.

By the time Jim turned nineteen in 1927, his muscular frame had filled out from six years of hard work in the mines, and he knew that he was a man — a man who could do as well, if not better, than anyone else in the mine, and decided to do something about it. So one morning he stopped Davie before his shift started.

"It's time I was in charge of my own room," he announced. "I can put up my own timbers and brattice."

"I've watched you for over five years now, laddie, and I know you're as good as your word. You'll need a partner."

"Bob Crawford can work as hard as me. We'll put out a lot of coal."

Davie had seen how Jim worked and he thought that, although Bob did not have the same determination, he might improve working with one who never quit.

"If you say you can handle it, then you can start digging next Monday."

"You've been good to me, Davie," Jim replied. "I'll show you what output is."

"Don't go killing yourself, lad. Just work steady at it."

Now Jim felt like a true miner, in charge of his own work area and paid for the amount of coal he could send out of his room each day. He persevered and quickly became one of the top miners — but the only thing different in his life was the money. He was still working in the belly of blackness, and the stirrings within him continued. *There has to be a better way to live than this.*

Meanwhile, he had become well-known in the whole region because of his enthusiasm for dancing. So when some in the local community were looking for people to run the Saturday night dances in the Castle Campbell Hall in Dollar, they approached Jim and his gang and asked them to do it. Without any hesitation, they jumped at the opportunity and, with the help of Geordie Bryce and some of his talented musician friends, the new venture was underway. It soon became a popular spot for young folk and the gang of seven began to make a little money on their enterprise.

The pattern on most Saturdays was for Jim and Bob to have a few beers "to wash down the dust" and then meet Mary Wilson and Bob's girlfriend, Min, and head off to these dances. Following each dance, they would catch the last bus from Dollar to Tillicoultry. If Jim was delayed in cleaning up after the dance and missed the bus, they would walk with some of the other stragglers and continue their lively banter along the footpath beside the river.

His sister Mary, now seventeen, asked if she could attend the Saturday night dances. Jim reluctantly agreed to allow her to come but made it clear that she was to go home with him. He was very protective of this beautiful young woman and cared for her with a fondness that went back to their early days on the paper route. The members of the gang knew how Jim felt and, although Mary was a real beauty, none of them ever tried anything funny with her, becoming watchdog older brothers instead. They loved to tease Jim about taking his two Marys home each week. For some time, Mary Wilson resigned herself to knowing that if she was to spend time alone with Jim, she would have to see his sister home first. Still, she did not like having to play chaperone and asked him if there might be another way for the girl to get home from the dances.

"But couldn't Geordie and Cis see her home?" she queried, snuggling up a little closer to Jim's side as they walked.

"I'm the one to do it, and that's it," Jim replied firmly.

One Saturday evening in January 1928, Jim did not want the dance to end. He had a great time and felt free as he glided around the floor — so much space, so much life — and, even with the smoke in the air, it was so much easier to breathe than his daily lot offered. Dancing with his sweetheart was such fun as long as they were in the crowd and moving, but anytime he thought of the walk home, he wished he could avoid it. Lately, the only topic on Mary's mind was getting married and raising a family. He thought he loved her but felt that marriage was not what he wanted right now. The relentless routine over six years of days and nights spent underground was not something he wanted to continue forever. Certainly he was now a full-fledged miner, doing well in some ways, but all he could think of was that there must be more. *I don't want to end up like my mother and father — a house full of children, little money to cover the expenses, hard, dirty work all day, every day.* When he tried to share these thoughts with Mary, she usually turned the conversation back to marriage and raising a family in Tillicoultry.

When the dance ended and the last bus had left, Jim began to clean up, putting off the walk home as long as possible. After they had dropped his sister off, Jim grew silent on the walk from Devonside to Tillicoultry.

"You're quiet tonight," Mary said.

"Och, I'm just tired and need my bed," Jim replied quickly.

"Well, you hardly missed a dance. I don't know how you do it."

"Aye, it's one of the few things that are good in my life."

"What about me?" Mary asked, tugging on his arm.

"You're part of it. I'm just tired tonight."

Mary accepted his word and was content just to walk arm-in-arm with the man she loved. When they reached her house, Jim gave her a quick goodnight.

"Well, I'm away home now, lass."

"Stay a little . . ."

"I'll see you in the kirk in the morning and then we can go for a walk up Alva Glen," he offered quickly as he turned and began to walk away.

"Jim, we need to talk some more," she pleaded. "What is going to happen to us if you keep thinking that you want more that you have now?"

"We'll talk tomorrow. I'm away now," he said abruptly.

He did not look back as he headed south toward Devonside.

Mary's words held fast to his brain as he walked home. *What's going to happen to us?* He paused on the bridge over the Devon, one of his favourite spots, took out his tobacco pouch and carefully rolled himself a smoke. He reached into his jacket pocket for his lighter, gave the wheel a sharp snap with his thumb, put the newly-lit flame to the cigarette and drew in deeply. He looked down at the Devon as he quietly enjoyed his smoke. There were many memories attached to that river. Fishing, swimming, family outings . . . and then Meg came to mind. She was doing well in Glasgow. Henry had served his time, been divorced, and had properly married Meg. They had another child coming and were pleased that her parents were raising John as their own. Jim reflected on how long and hard the road was for Meg, but she was happy with her new life.

She was doing what she wanted to do. Again, the words she had spoken to him when he had made his decision to go into the mine rang vividly in the ears of his memory.

"That's no life for a smart lad like you. Don't waste your life."

She was right. I want more.

He had already been thinking about leaving and making a new start somewhere like Australia or Canada, where he understood that there were more jobs than workers.

Although his dad was working less each year and his mother had come to rely on him for moral as well as financial support, he knew that the others could pick up the slack if he were to leave. Cis had married George Bryce and was living in Alva. Doll was going with John Cairns from Fishcross and would be close to home. Mary was working in the mill and had begun dating but still helped her mother. Jock and Bill were working in the mines and bringing in money to help feed the younger children, Robert, David and John.

Gripping his cigarette between the first two fingers of his right hand, he placed it tightly to his lips and inhaled the long final drag until the burning tobacco almost touched his fingers. He held the smoke in his lungs for a moment and flicked the butt into the river below. He reached into his pocket, and fingered the money that had been his share of the profits from the dance.

What if I could save enough . . . ?

• • •

They had two hours to kill before the bus would take them home after their painful three-nil loss to the Stirling Albion football club. As they strolled about town, their usual fish and chips did not bring them the usual comfort.

"Bob, look at this," Jim exclaimed as they were passing the window of the Canadian Pacific Steamship agency in Stirling.

Jim pointed to a poster depicting a peaceful rural scene with a fenced yard full of farm animals in the foreground, a rolling field with stooked bales in the background, and the smiling face of a handsome farmer admiring the scene. Superimposed on top of it all was the invitation, 'Come to Canada — Farm Workers Needed'.

"What do we know about farming?" Bob asked.

"What did we know about mining before we started? Let's go and ask about it."

"You mean now?"

"Aye, now."

He pointed to a note by the sign that read 'Apply within for full particulars.'

"Are you coming?"

Bob followed Jim inside.

"Excuse me, sir," Jim said to the man behind the desk. "We'd like more information about Canada."

The man slid his glasses down his nose and peered out from under his leather visor.

"Well," he said, "I can give you an application form. Fill it in and send it to the Glasgow address. You'll find out in a few weeks if you're accepted."

"Can you tell us how it works?" Jim asked, not wanting to leave without knowing more.

The agent paused, looked from one lad to the other, then launched into an oft-repeated monologue on how the government of Canada paid Canadian Pacific to recruit farm workers and process their immigration forms. He went on to explain that the immigrants were required to pay a small amount to the company for transportation and then would be assigned to a farm and agree to work there for two years.

"What are you working at now?" he asked, finally.

"The Alloa Coal Company," Jim replied, raising himself to his full five-foot-seven-and-a-half inches, "and we're both hard workers."

The agent moved to a well-organized rack on the back wall, retrieved some papers and offered them to the young men.

"Here are your forms."

Martha and Jimmy were extremely upset when Jim returned from his football trip waving a Canadian immigration application form. It was one thing to have their son talk from time to time about going to Canada or Australia, but they had not given the idea much credence, especially in the past few months as he and Mary Wilson seemed to be getting more serious. Nature would have its way and soon he would settle down as the others were and the foolishness would stop.

"What do you plan to do with that?" Jimmy demanded.

"What do you think I'm going to do?" Jim retorted in his usual cocky manner. "I'm going to Canada."

"You're soft in the head," his father blurted. "You can't be serious about going so far away."

"You've not enough experience to head off by yourself," Martha interjected. "Dinnae be daft."

"I'm not daft. I know exactly what I want to do," Jim insisted. "I'm going to make a better life for myself than I can ever have here."

"I know that when you get your mind on something, there is no changing it," Martha said, "but do you have to leave the country? There must be work closer . . . like in Glasgow?"

"Mother, we've been over this many times," Jim replied, his voice beginning to express his exasperation. "Without a trade, there is nothing for me except the mines. I won't spend the rest of my life in them. Nor will I raise a family to work in them. Can you not hear me?"

"But you will be so far away, and you're such a help to me. I depend on you. Maybe your Aunt Liz will be able to . . ."

"Don't bring Aunt Liz into this!" Jim interrupted sharply.

He choked back the frustration that was triggered lately whenever her name had come into conversations. Many a winter's night as he trudged home from a hard shift in the mine, his damp clothes grabbing the chill of the air and thrusting it into his body, he had thought about what it might have been like for him if he had gone to Glasgow instead of the mines. It was too late to bring her back into the picture.

"Can't you be happy for me? I've finally got another chance to better myself."

"But going so far away will be hard on your mother and me," Jimmy said, knowing that the more upset Martha became, the harder she would come down on him. "Look how you've upset her!"

"I'm sending this application in."

With that, Jim turned away and walked over to the table. He sat down, spread the forms in front of him and began to read through them.

"Looks like I'll need some dates and places from you."

It took two more days of serious conversation before his folks reluctantly provided him with their background information and the forms were completed and mailed.

The following Sunday afternoon, under the warming March sun, Jim and Mary Wilson strolled up the burn to Alva Glen.

"You're awfully quiet today, Jim," she remarked.

"Aye, I've a lot on my mind these days."

"Is it about us?" Mary asked eagerly.

For the past few months, she had been hinting at marriage and raising a family. Now she was hoping that Jim was ready to talk seriously about it.

"Aye."

He took off his coat and spread it on a warm rock at the stream's edge.

"Let's sit awhile."

She sat beside him, reached out for his hand and asked, "Are you all right?"

"I've got some good news. I'm going to Canada."

"You're what?" Mary exclaimed.

"Bob and I are going to Canada to work on a farm and make money," Jim enthused.

"What about me? And what about Min? I thought you wanted to get married."

"I can't do that right now, lass," he retorted. "I don't want to have the kind of life my family has, bringing bairns into the world with never enough to eat or decent clothes or a proper house."

"Jim, I love you," she said softly, tears rolling down her cheeks. "We'll manage fine. I don't want you to go."

Jim pulled her close.

"When I've made enough, I'll come back for you. We can get married and go back to Canada."

"You know I can't leave my mother."

"Your sisters can look after her."

He paused, his eyes open but not taking in the burbling stream. *It's now or never*, he thought.

"Mary, I have to go."

"Why can't you be like everyone else?" she sobbed.

"That's just it. I don't want to be like everyone else. That's not who I am."

He had no more words, but she had tears not yet shed. The two of them sat in a stillness broken only by her soft crying and the burbling burn. The silence continued as they wound their way down the glen to Mary's home and it remained between them for most of their time together during the anxious days of waiting for word from Glasgow.

This was also a very tense time for Jim at home, although Doll was supportive of his venture and encouraged him to be patient as he awaited the reply. His sister Mary, a bit of a romantic, talked of when she might have enough saved to come and see him in the new land. Martha expressed her despair with angry outbursts at whoever was close at hand. Jimmy found solace with his friends and in the pub.

At least the walk with Bob to and from the mine each day was much more spirited and gave Jim an opportunity to go on at length about what he hoped might be their fate in Canada.

"There can't be any work on a farm that is as hard as digging coal," he gushed. "We won't get dirty or cold and we'll be making good money and eating well."

"Aye, it will be good," Bob replied with little enthusiasm.

"What's the matter, lad?"

"Och, I'm a wee bit tired, I guess," Bob answered. "Min and I were up late again last night."

"You've got to cut back on that courting or you won't have any life left for the boat," Jim gave Bob a nudge with his elbow.

"I'll do fine," Bob replied flatly.

Jim looked at his pal, noticing he seemed preoccupied with something.

"Well, I'm hoping we don't have to wait much longer for word about whether we'll be going or not."

Finally, a letter arrived notifying Jim that his application had been accepted. All he had to do now was remit the fee of £3.10d to cover the cost of transportation from Greenock to Quebec City by steamship, and from there to Edmonton, Alberta, by train.

"So you're going ahead with this?" Martha interrupted as Jim re-read the letter for the third time.

"You know I am, Mother," he bubbled. "Isn't it wonderful that I am going to get a chance to see more of the world?"

He found the map of Canada that the agent in Stirling had given him and, after much searching, found Edmonton.

"It's almost at the far end of the country!" he exclaimed.

Martha was not about to get swept up in his enthusiasm and tried to overshadow his sunshine with black clouds of doubt.

"You don't know how to care for yourself, laddie. We've done everything for you. You'll get into trouble. You'll get hurt. You'll be so far away we will never see you again."

Jim had become inured to her constant challenges and ignored her protests.

"After supper, I'm going to see if Bob got his papers, too."

"Are his parents happy about him going off to another country far away?" Martha challenged.

"They want to see him do better for himself," Jim retorted.

Later, he called by Bob's home. His pal came out, a deflated expression on his face.

"I guess you got your letter too, eh?"

"Aye," Jim said, paying his pal's dejection no mind. "I did and I want to send the money as quickly as possible."

"Let's go down by the burn, Jim," Bob said, leading the way. "I have something to tell you."

Jim took notice of Bob's expression and sombre tone.

"What's up? You're looking fair worried."

"I canna go, Jim."

"What's the matter?"

"Min's knocked-up."

"Bloody hell! What are you going to do?"

Farewell to Scotland

"I'll have to marry her, Jim."

Jim kicked a rock as hard as he could toward the stream.

"That's a bugger."

"What will you do?" Bob asked. "Will you go by yourself?"

"You're damn tootin' I will, come hell or high water. I don't want to get trapped here."

"I wish I was going with you," Bob said sadly, "but I have to do what I have to do."

"You know that and so do I. But it's all the more reason for me to go now," Jim added.

He turned from his friend to scuff another rock and hide the tear that had suddenly formed.

The official letter arrived in late April and informed Jim that he would be sailing on May 12, 1928, on the Canadian Pacific Steamship *Montclare* from Greenock, the port just outside of Glasgow. He had about three weeks to get ready — to purchase a trunk and some new clothes and organize his few possessions for shipment to his new land. He quit his job so that he could prepare properly. It was a time of goodbyes, some harder than others.

His pals were happy that one of their own was setting out on an adventure and they posted notices in Dollar, Tillicoultry, Devonside, Coalsnaughton, Alloa and Alva, inviting all Jim's friends to a farewell party on May 5th, at the Castle. And what a party it was! Geordie recruited some of his buddies to supply music, some of the young women organized a lunch and the young men looked after the liquid refreshments.

Jim did not miss a dance all night and was glad of the lunch break just before midnight that helped refuel him. Mary Wilson sat with him while he ate sandwiches and quaffed ale.

"Jim," she said, "I know this is your party, but I've not seen much of you tonight. Everyone wants a dance with you."

"I know, hen," he said.

He put his glass down and took her hand.

"I've no choice," he told her, "but to see as many friends as I can tonight. We'll have our time later, and, with me no longer working, there will be many hours for the two of us this week. Can you give me a smile?"

"It's not easy. Will you at least have the next dance with me?" she pleaded.

"Aye, and one or two more before this night is over. Come on, Geordie's warming up his squeezebox."

Following one last gulp of ale, he led his sweetheart onto the floor.

The partying went on long after the last bus had left Dollar but, on this night of celebration, the dark path was brilliantly lit for the revellers by a moon only a night away from being full.

Martha became more disagreeable each day as the departure of her eldest son drew near. She bombarded him with lectures about how to take care of himself, how important it was to stay away from strange men and how difficult it would be for him to stay out of trouble in a foreign country. Jim tried to assure her that she would manage fine.

"I'll send you money as soon as I land a job."

• • •

She refused to see beyond her present hurting.

Jimmy quietly supported his son and encouraged him in subtle ways, so as not to upset Martha. In letters to his sister Liz and his brother Robert in Glasgow, he arranged for them to meet the bus and take him and Jim out to Greenock on the day of departure. Without revealing his own need for contact, he urged Jim to write to his mother regularly.

"She'll need to know that you are all right."

Jim had heard that there was a fellow by the name of Dunc Heddleston in Alva, two miles away, who would be sailing on the same ship. When they met in Tillicoultry, they were pleased to discover that both of them were headed to farms in the Edmonton area. They made plans to meet on the boat in Greenock. Jim volunteered to write to the office in Glasgow asking if they could room together.

It was hard saying goodbye to his sweetheart on his last night in Scotland. Mary cried and held on to him so tightly that for a moment he wondered if he was making a mistake. But he knew that there was no turning back now and it was time to end this lingering farewell.

"I have to go now, Mary," he said abruptly, stepping back from her clinging arms. "You know I don't want to hurt you, but I have to spend some time at home tonight for my mother's sake."

"Don't go. Don't go," Mary pleaded.

"I'm not going to talk about it again. Can we just say our goodbyes and get on with it? Will you be at the bus in the morning?"

"I'll look like a daft fool if I go. I won't be able to stop crying," she said through tears she believed would still be flowing the next day.

"I love you, lass. I'll write. You know '*I'm no awa' tae bide awa*,'" he half sang the words of the old song. It was always his way to shatter the tenderest moments with some form of levity.

On his way home, he stopped on the Devon Bridge where he had often paused for a smoke. Now his practiced, nicotine-stained fingers slowly fashioned a cigarette and placed it between his lips. He pulled out his lighter and, with his hands forming a protective shield around the tiny flame, he lit his creation. His body responded with a long drag, he held the smoke in his lungs, savouring the warm, calming feeling, then slowly exhaled. The past month had been a whir, so much happening, drastic changes: sad goodbyes, exciting possibilities, tears, celebrations, and now he faced his last night at home. After one last long pull on the cigarette, he flicked the butt into the river and continued walking.

Doll and Mary were sitting at the table drinking tea with their parents when Jim arrived home. The boys were all in bed as Jock and Bill had to work early in the morning and the others were too young to stay up late.

"Would you like a cup of tea and a biscuit?" Doll asked, rising to get the teapot without waiting for a reply.

"Aye," Jim said.

Doll filled a cup for him. He reached out for a biscuit and poured a little treacle into a dish so he could have his favourite sop.

"How's Mary?" asked his sister of the same name.

"She's all right. Does a lot of crying," Jim said quietly.

"Tears don't bother you, lad, do they?" Martha jabbed with sarcasm.

"Mother, don't start on him tonight," pleaded Doll. "It's our last time together for God knows how long."

"We'll have to be up early," Jimmy declared as he drained his cup. "The bus leaves Murray Square at a quarter of eight and will not wait for us. Dolly, fetch the tea."

Following some quiet conversation about last-minute checks on Jim's packing and reminders to say hello to the Glasgow aunts and uncles, the family drifted off to bed.

Money being what it was in their home, it has been decided that only Jimmy would accompany his son to the ship. As the family watched the bus pull out early on that Saturday morning, Martha clung to Doll while the two tearful Marys stood arm-in-arm, waving. Jim twisted around in his seat to watch the little group quickly growing smaller and finally disappearing.

"The crying's hard for me to see," Jim said, turning to his father. "You'll no' greet, will you?"

"If I do, you won't see it, so don't worry, laddie."

The bus turned onto the main road and headed toward Stirling, with its dominant castle. Jimmy pretended not to see it.

"Let me tell you," he said, "about the time William Wallace came to Stirling . . ."

It was good for Jim to have his aunts and uncles on the dock with his father on that late Saturday afternoon in May as he prepared to take his leave of home and country. They were in good spirits and wished him well as he set out. He had the required £10 landing money, which was to cover his food costs on the train to Edmonton, but before he went up the ramp to board the ship, he turned and handed money to his father.

"Father, here is £5 for my mother."

"Jim, you will need that money for food in Canada," replied Jimmy.

"No, I'll be fine. I want my mother to have this."

With that he shoved the bill into his father's pocket.

"Well, I'm no *awa' tae bide awa'*. I'll come back and see you when I'm rich."

With a hearty handshake, he left his father and proceeded to the tourist-class loading area to look for Dunc.

Chapter Nine
The Alberta Coal Branch
January 1928

"There's a new mine opening at Mile 40," Frank announced one day in, looking at his son-in-law. "I think we should try it. What do you think, Carl?"

Carl looked across the table at his wife and before he could answer, she did.

"You always want to move, Papa." Mary said. "What's so special this time?"

"The mine at Mile 40 is run better and they are paying more. The company has built forty new houses. Carl?"

"Well, if you think it is a good idea." Carl, a timid man who went along with whatever his bride said, was really waiting for her response. "Is that okay, Mary?"

"If it means we have more to spend and I can be close to Mama, what's one more move?"

A few weeks later when the family boarded the train in Mountain Park on their way to their new home along the coal branch, they were heading to familiar territory. They had been living about ten miles west of Mile 40 at the Yellowhead camp when Ida was born.

"There is no school here!" Ida announced, following her exploratory walk around the community that centred on the mine and its company store and post office.

"What are we going to do, Papa?"

"The pit boss told me all the kids here take correspondence."

"But I like school. It's not fair!"

"When you are the one bringing home the money, then you can decide what is fair. Now help your mother."

The closest village with schools and other services was the junction town of Coalspur, five miles back along the rail line, but Frank had decided that Mile 40 was the place to live in spite of all its limitations for his family. However, he did care about his children's education. He was frustrated because his English was limited and he could not utilize the extensive education he had in Italy, but was determined his children would do better. They would learn the language and be able to get ahead just as the Englishers' children did, so now he enrolled them in correspondence courses and pushed them hard to maintain good grades. He forced the four of them to sit around the table every day. After work each night, he helped them prepare their lessons for the weekly mailing to Edmonton. The children worked hard at their lessons and even harder at keeping their feelings toward their tyrannical father buried during that long winter.

One morning, enjoying a much-welcomed break from the relentless schoolwork, Ida and her younger sister Stella slowly trudged home through the snow with milk for the next day's breakfast. They talked about how awful it was living at Mile 40. Ida disliked her father for the way he always decided to move without ever asking anybody what the family thought about the idea. As they walked, the snow from an open field swirled across the road in front of them and settled gently into the drift that was forming along the fence.

"If he only knew what I was thinking, he would kill me," Ida said.

"Ida, don't . . ."

"He is a mean man and he doesn't care about us."

"Don't talk like that! Don't even think like that!" Stella cautioned.

"Well, look at the way he hollers at Mama all the time. I think he slaps her too when we're not around."

"Stop it, Ida." Stella began to cry. "I don't want to hear stuff like that."

"Don't be such a baby. I would never say what I think . . . to *him*. Stop crying before we get home." Ida grabbed the end of Stella's scarf and wiped the icy tears from her sister's face. "I always get into trouble when you cry."

Stella looked up at her older sister and squeezed the last tears from her blue eyes.

"It's no wonder everyone calls you a rebel," she said. "You don't know how to keep quiet."

In spite of all the negative energy generated around the kitchen table, there were some pleasant evenings when Frank would read science fiction novels to his children. In order to continue learning as much English as he could, he always read with an Italian/English dictionary beside him, and would pause to look up words that puzzled him. The children especially liked Edgar Rice Burroughs' *The Land That Time Forgot* and others in the series about the land of Caspak. Angelina did not participate, still refusing to learn the language.

At other times, Frank would wind up the phonograph and carefully place one of his precious opera records on the turntable. He loved opera and one of his favourite stories was of the trip to Milan in 1906, when he was twenty-one years old, to attend the performance of Verdi's *Nabucco* (short for *Nabucodonosor*) at La Scala.

"You are getting older now," Frank said, one evening as he told the family to sit down and listen to the opera. "The man who wrote this opera, Verdi, was very clever, and told the story of the Jewish people when they were ruled by King Nebuchadnezzar of Babylon. *Nabucco* is based on the story of the Babylonian king and tells the story of the Jews when they were exiled and ruled by other people. Verdi wrote it in such a way that it was also the story of Italians under the Austrians. When the second record had finished just after the powerful 'Va, Pensiero', the song the Hebrew slaves sing as they long for their homeland and freedom, Francesco paused to translate for his children, who understood Furlan, but not Italian.

"Listen to the words," he urged. "'*Oh, my country so beautiful and lost! Oh, memory so dear and fatal!*' That stirred up the people so much that the Austrian government forbade any encores at the end of an operatic performance. However, the audiences at La Scala demanded repeats and the conductors decided a disappointed audience could be more dangerous than the Austrians and, in an act of defiance against the régime, gave the encores."

"Did they get into trouble, Papa?" Ida asked.

"No, the Austrians were wise enough to know that there might be a rebellion if they did. But, what they didn't know was that when people shouted 'Viva Verdi' as they stood at the end of the performance they were really shouting praise for their king."

"I don't understand what you mean," Ida interjected.

"The name V-E-R-D-I, became an acronym for *Vittorio Emanuele, Re d'italia*, king of Sardinia who later was crowned the first king of the united Italy. Now that you know this, let us listen to the rest."

Although the children were not enthralled by opera, they knew that with the records playing, their father would not be demanding anything of them.

When the school term ended that year, all four children walked the five miles to Coalspur to sit their exams. In spite of their resentment at the circumstances of their lives, and their anger at their father for his inconsiderate decision-making, Pete and Ida were both able to maintain their honours standing that year when he finished grade nine and she, grade eight. The grades meant nothing to Frank, who had decided that this was as far as they would go in school.

"But, Papa!" Ida exclaimed. "I like school and . . ."

"Enough! You have enough for a woman and we can't afford to send Pete off to high school in Edson."

"It's not fair . . ."

"Fair? What do you know about fair? Those Englishers think I'm stupid because I don't understand all the words. I'm one of the hardest workers they have but I never get the best cuts."

"You've done okay, Frank," Angelina soothed.

"Don't be stupid. I could do better if I were English. Where's my supper?"

The room was quiet as the family settled in to eat. None of them wanted to be the first to speak and thus incur the wrath of their father when he'd had one of *those* days in the mine.

Eventually sensing Frank was calming down near the end of the meal, Stella decided the time was right to make a request.

"Can we play poker tonight, papa?"

"Yes, as soon as you get the dishes done."

The family often played poker for fun in the long winter evenings, sitting around the table dimly lit by the 25-watt bulb hanging from the ceiling. Frank had taught the children the basics of card values and the fine art of bluffing and betting. Prior to playing, each player was given a quantity of chips and the object of the game was to end up with more than any of the other players.

"Papa," Stella called out, "the table is cleared and we are all ready for a game."

The game went well for awhile with chips being won and lost.

"She was bluffing," Pete blurted out after one hand, as Ida swept the chips that she had just won, to her pile. "She only had a pair of fives."

"That's was a smart play, Ida" Frank said.

He shuffled the deck and dealt a new hand.

"Watch me this time," said Pete.

He looked at his cards and glared across the table at his sister.

"Hey Pete," Stella said. "It's only a game. Try and enjoy it."

Pete did not like to lose, and in his determination to show the family how good he was, overbid his next hand. When he looked down at his diminished chip supply, he jumped up and scattered the chips.

"I don't want to play this stupid game anymore," he cried.

Frank flew into a rage, grabbed his son and threw him to the floor. Sitting on top of him, Frank reached out for the coal shovel with which to spank the obnoxious fourteen-year-old.

"Stop, Frank! You'll hurt him," Angelina pleaded, knowing how quickly he could become violent.

"Shut up!"

"He's going to kill Pete!" Stella screamed.

"No, he won't!" yelled Ida.

She leapt onto her father's back, trying to pry him loose from her brother. Soon, Stella and Johnny were into the fray while their mother looked on, crying helplessly.

"Leave him alone, Papa!" Ida commanded as she tore at his shirt collar.

The loud zipper-like *ri-i-i-p* as the shirt gave way stopped the action as though it was a movie still. No one spoke. The pile separated. Angelina led her panting husband to the bedroom as Pete and Ida cleaned up the mess and encouraged the other two to get off to bed.

Frank said nothing more about the incident but the tension remained, with each family member determined that he or she would not be the cause of the next outburst. Because he felt he had no power other than at home, he continued to control the family as much as he could. Ida, like her mother, stifled her negative feelings toward Frank, while Stella and Johnny tried to look at everything as fun. Pete began to spend more of his free time visiting with Mary and Carl, away from the pressure of his father's temper.

Frank's sister, Therese, had moved to Vancouver and visited the family during the summer. It was good for all of them to hear about relatives the children had never met. She brought a gold chain that Umberto had cut into lengths so that each of his children could have a share. Therese gave Frank his portion of the heirloom, which he accepted without showing any emotion. They did not talk much about their father during her visit.

Angelina did well with the meagre resources available during those hard Depression years, feeding her family with nourishing and filling meals. Though in time she had become a good cook, she could never meet Frank's expectations. His family had always eaten more elaborately than had hers, and he constantly put her down for her lack of finesse in the kitchen. He did cook when they were hosting a family gathering or having a special celebration and made sure that Angelina and the rest of the family knew that he was only doing it to make sure it was done right. She suffered his humiliating comments in silence and, because she felt inadequate as a cook, did not share much of her kitchen experience with the children. The girls had to help with the cleaning, the washing and ironing, but not the cooking.

The one exception came when Angelina was making polenta in her special polenta pot. There was a point in the process where it needed two people so, when Frank was working, she would ask Ida to help her with the corn meal. Ida's job was to pour the corn meal slowly into the pot of boiling water while her mother stirred and stirred. She did not trust anyone else to stir it right and did not want her pot burned because of inadequate stirring. She was able to keep the polenta from sticking and, at the same time, stir the spicy tomato sauce mixture in another pot as the meal preparation progressed. On polenta nights, Angelina would cook chicken, make the sauce, and combine them to simmer until the spices melded in a mouth-watering blend, whose aroma tantalized the family. Then the table ritual would begin.

"Get the board ready, Ida," said Angelina. "The polenta is almost done."

Ida got out the well-oiled board that looked like an over-sized breadboard and placed it on top of the oilcloth table cover. She then arranged a fork at each place and all was set for dinner.

"It's all ready, Mama," Ida said.

She and the other children would impatiently wait for the completion of the polenta ritual.

Frank picked up the polenta pot, carried it to the table and holding it over the board, tipped it so that Angelina could take her wooden spoon and spread the steaming yellow mass onto the board. Frank then picked up the pot of bubbling sauce and slowly poured it over the polenta while Angelina took her wooden spoon and spread the chicken pieces evenly.

Polenta was one of the family's favourite meals because it was not only tasty but, since each person ate from the board, there were no dishes to wash afterward. When the supper was finished, Angelina would carefully cut around the untouched portion, removing the rim from which people had eaten. The remaining polenta would be stored for lunch the next day.

As the Linteris family moved around the Coal Branch, Angelina continued to welcome the periodic visits of the itinerant priest, Father

Louis. Although she was shy about conversing with him in her halting English, and afraid that her cooking might not be good enough for this wonderful man, she always invited him to stay for a meal with the family. When she heard that Father Louis was in town, she would send Pete out to catch some fresh trout, the priest's favourite. She would make an oven-load of fresh bread, pan-fry the trout in bacon grease and top off the meal with vegetables from the garden.

Father Louis was a kindly man and fond of the well-mannered Linteris children, especially Ida, who appeared to be the sensible one of the two daughters. So, without saying anything to the family, he contacted his bishop and suggested that he knew of a girl who might become a nun.

It was quite a surprise later in the year when the bishop was travelling through the area confirming people, to have him drop in at the Linteris home.

"Will you have a cup of tea?" Angelina timidly asked her esteemed guest after he sat down at the table.

"No, thank you. I have many people to see today and cannot drink too much tea." Then he came right to the point of his visit. "I understand you have a daughter."

Frank, assuming his role as head of the house, replied, "We have two daughters at home, Stella and Ida."

Looking at Ida, obviously the older of the two, the bishop asked, "Which one are you and how old are you?"

"I am Ida and I am almost fifteen, your grace," she replied quickly.

She looked away from his piercing eyes toward her father, unable to express in words the fear that had suddenly gripped her.

"Your daughter," the bishop continued, turning to Frank, "is a fine young woman and appears to be just what we are looking for in the church. Would you agree to her joining a convent and taking her training to be a nun?"

Frank could see the obvious discomfort in his daughter's face. When he had gone to mandatory confession prior to his wedding in Italy and had unloaded what he felt the priest wanted to hear, he had decided it would be the last time he ever did that. He did not need to tell his sins to another man.

"I think you should ask her that yourself."

"Well, Ida, what do you think? Would you like to come to Edmonton and become a nun?"

Her feelings toward nuns had not changed from her earlier school days. Knowing her father wanted nothing to do with the church and would not object, she decided to answer straightforwardly.

"No, thank you. I have other plans for my life."

Frank said nothing, but his nodding half-smile spoke volumes about his feelings toward the church and his daughter's decision.

The visit had not achieved its purpose so the upset bishop stood up abruptly, thanked the family for being so kind to Father Louis and took his leave.

Chapter Ten
Dreams Unfulfilled

The *Montclare* and her sister ships — all of them built on the Clyde — had undergone a refit just a month before this May 1928 sailing. Many of the underused first-class cabins or 'cabin class', as they were called, had been converted to tourist-class. While this had doubled the number of people that could be carried, the facilities were still first-rate, as was their location on the ship, on the deck just below the cabin-class passengers. The third-class passengers, most of who came from Poland and the Ukraine, were in the lower decks, crowded into sleeping dorms with hammocks, and were provided a separate gathering place and a dining room.

A uniformed steward showed Jim and Dunc to their tourist-class cabin, explaining that they would be sharing it with two other young men.

"My God, look at this, Dunc!" Jim exclaimed. "Four bunk beds, two desks and a toilet . . . all of this for just four of us."

"Aye, it's grand," Dunc offered. "And we have a little window to look at the sea."

Jim climbed onto the top bunk nearest the porthole.

"Let's pick our beds before the other two get here," he suggested. "I'm not having anyone sleep over me."

Taking a lead from his new friend, Dunc clambered into the other upper and the two lay back to savour their new-found luxury.

A short time later, the steward reappeared with their roommates, Doug and Angus, friends who had worked together in a rubber factory in Glasgow. They too had responded to a Canadian Pacific poster and dreamed of farming.

When the newcomers had been introduced and claimed their lower bunks, the steward asked for their attention.

"You lads can unpack later. It's time for supper. I'll show you where the dining room is."

They followed him along the corridor and up a flight of stairs into a large room filled with people milling around linen-covered tables.

"What a layout!" Jim exclaimed.

He had never seen the like before, not even on his visit to Stirling Castle: the opulence of the highly-polished wood paneling, the sparkling chandeliers with crystal pendants splitting the light into millions of sparkling diamonds, the elegant tables with their own display, as reflections bounced off the crystal glasses and the lustrous silverware. Another steward, a Scotsman, showed the four gawking young men to a table. As they sat down, Jim looked at his mates.

"Look at all the fancy silverware," he mused. "What will we do with that lot?"

The steward smiled at the group, obviously first-time travellers, handling them menus and discreetly suggesting the order of use for the cutlery.

"If you have any other questions, just call me over. *Bon appétit.*"

His parting words were reflected in the menu, where half of the words were in French. Four pair of eyes scanned the menu, moving from one side to the other and then, almost on signal, they looked up, each waiting for some sign of comprehension from another. Angus caught the eye of the steward and beckoned him over.

"Could you gie us a hand?"

The smiling steward responded quickly, explaining that Canadian Pacific printed menus in French and English on their ships, trains and hotels, as both languages were spoken in Canada. He pointed out that the French side was exactly the same as the English, and asked if he might make any suggestions. The very first item was a puzzle, as none of them knew what *hors d'oeuvres* meant. If this was the English side, it didn't look like English. With practiced patience, the steward described each item on the menu and soon they made their selections. Course by course, as the waiters served the sumptuous meal, the young men devoured each offering with gusto and joy.

Jim took a roll from the basket and began wiping every bit of gravy from his plate.

"I don't think they do that here," Dunc whispered.

"What? Did you not learn 'waste not, want not'?" Jim replied with some incredulity. "We'd get a tongue-lashing from our mother if we ever left any food on the plate. Besides, it's the best part of the meal."

Trying to hide his embarrassment, Jim's shy friend forced a smile as he glanced at his other tablemates. Nobody said a word as Angus slowly put a freshly broken roll back into the basket.

When they had finished the chocolate dessert and washed it down with strong tea, Jim still wished he could have had one more roll with jam on it, his usual way of ending a meal.

Touring the ship following supper, Jim told Dunc, "That was the best meal I've ever eaten in my life. I wonder what they'll give us for breakfast."

They stood quietly for some time on the deck that evening taking in the expanse of ocean, still well-lit by the waning moon, now in its last quarter. Taking in the scene, Jim thought, *The farther I get from home, the blacker it becomes. The brightness of my life with Mary and my friends will become dim. Burns would have written a poem on this.*

"Do you have a sweetheart, Dunc?"

"No, there weren't a lot of choices in Alva, and I was usually tired after a long day at the mill. Do you?"

"Aye, her name's Mary. Didn't want me to leave. I had to promise I'd come back and get her when I get rich in Canada."

"That could take a while. They don't pay that much on the farms."

"She'll wait. Wants to stay by her mother as long as possible. But I'm a hard worker and will make money faster than most."

Jim was up early in the morning, continuing his exploration of the ship as he waited for seven o'clock and breakfast. Walking from one end of the *Montclare* to the other, he figured, was about as long as two football pitches. It was not too crowded at that time of day and he could strut out for a measurement.

Following his first breakfast on board, when he consumed more food "than I usually eat in a month of Sundays," Jim went to the smoking room for his usual after-the-meal smoke, with Dunc in tow. Later, they moved on to the writing room, where they wrote postcards home.

Jim informed his mother and Mary of his great accommodation, his new friends and even detailed as many of the menu choices as he could fit into the space. In future cards, he would tell of what it was like browsing the massive library and touring the decks.

As the journey continued, Jim's cabin mates marvelled at his ability to eat and enjoyed his never-ending comments on the wonders of the grub. If he said it once, he said it a hundred times during that week: "These are the greatest meals I have ever seen in my life." Food had always been a priority, but he had never had this variety, quality and quantity, so each meal was a gastronomic adventure for him and his friends. However, he did most of the talking about food.

Of the many activities offered on the ship, Jim was drawn to the dances, the professional entertainment and the concerts in which passengers participated. By the second night, he was on the program in the cabin-class lounge imitating one of his favourite entertainers, Harry Lauder. Many of the Canadians on the *Montclare* were familiar with this singing comic and when Jim began to belt out "A Wee Deoch an' Doris", they applauded and asked him to repeat it. Many of the people were intrigued by his broad Scots vernacular, especially as he exaggerated the rolling of his Rs, and the guttural 'heuch' when he came to the last line of the chorus, "*If you can say, 'It's a braw bricht moonlicht nicht' yer a'richt ye ken.*'"

Jim didn't need much encouragement to continue and did so with "Stop Yer Tickling, Jock!", "Roamin' in the Gloamin'", "I Love a Lassie" and "The Saftest O' the Family". People showed their appreciation by offering him all the alcohol he wanted, but neither Jim nor Dunc were heavy drinkers so they stuck with a few beers and tailor-made cigarettes.

One morning, Jim was hurriedly dressing in order to get to the dining room by seven. Dunc, who was feeling unwell, watched him enviously.

"Doesn't your stomach ever bother you?"

"I won't let it," Jim replied. "From the time I saw that first meal I knew I wouldn't get seasick on this trip. There's too much food to keep down."

With Jim's upbringing, there was no way he was going to waste any available food by regurgitating it.

"Will you hurry and get out of here?" the usually mild Dunc ordered.

He just needed to roll over, lie still, and not hear about food or *he* would vomit.

On their fifth day, they could see the giant rock that is Newfoundland in the distance on the starboard side. Jim knew from talking to crew members that once they were past this distant British colony, they would be able to see Canada. Standing at the rail that afternoon, he did a shift in his thinking — he no longer saw himself going farther and farther away from home but sensed that now he was nearing something. *Just a few more days and I'll be walking on a new land. Fresh air, sunshine and more good grub . . .*

But shortly after midnight on that fifth day a groaning, creaking tremor shuddered along the ship.

"Jim, you awake?"

"Yeah, what's that?"

"Damned if I know."

"Turn on a light!"

The four cabin mates hit the floor almost simultaneously.

"I was dreaming I was in the pit," Jim exclaimed. "In a cave-in."

"Something's wrong," Doug hollered. "Let's get out of here."

The first to get into his clothes, Dunc was peering out of the porthole into the black night.

"Can't see anything," he announced.

"Come on, let's go!" Angus hollered.

He threw the door open and dashed into the noisy hallway.

With fear pushing curiosity far into the background, a growing mass of people jostled each other up the stairs and into the fresh air of the dark night.

"What happened?"

"I don't know. Iceberg?"

Iceberg! The word reverberated among the growing throng, and the cold terror that followed was triggered by the unspoken word '*Titanic*' with its pictures from schoolbooks and yellowed headlines, "*1700 Perish as 'Invincible' Ship Sinks*".

"We're stopping."

"I wonder if it made a hole in the ship."

"Seven years in the mines and I never got blown up. Will I drown instead?"

"Please keep calm," an announcement bellowed from the speakers scattered throughout the ship. "Everything is fine. Please stay calm. We had a bit of a rub with an iceberg."

The cacophony from the milling crush of passengers stopped as though turned off by a switch.

"The ship appears unharmed," the announcement continued. "Please extinguish all lights and maintain silence on the outer decks to assist the night watch. Please obey all directions from the crew. We will wait until daybreak to get underway again."

An eerie hush came over the ship as the engines slowed and all exterior lights were turned off to give the peering crew as much help as possible in their search for the telltale phosphorescence that appears when waves lick the icy white mass of an iceberg. It was an uncanny feeling to be afloat on this huge ocean with the quiet throb of barely-running engines operating only enough to keep the power generators working. Jim, like the rest of the passengers, could see nothing but stars, but everyone listened intently as they waited for the faint echo that the incessant foghorn would produce if its moan hit a solid object.

"Look toward the stern," someone whispered. "I can see something."

Jim glimpsed the slight glow on the waterline of the fading ice mountain as the drifting ship moved on.

"Dunc," he whispered, "let's make our way along to be closer to the lifeboats."

"Aye. I was thinking the same thing."

There was no sleeping for the folk on board that night. People put their arms around loved ones, comforting those who felt they were doomed to drown. Others mumbled supplications to a God they hoped was listening. The *Montclare* was one of the best watched-over ships on the high seas as passengers huddled on deck. When the sun broke the horizon at 4:40, most paid little heed to the glorious day that was being presented.

Jim was able to work his way along the deck to watch as a diver was assisted into a bulky suit. Two sailors lifted the headpiece and secured it to the brass collar on the suit and then they connected a long black hose that would be his lifeline while below the surface of the water. Tools hung from his belt as he was slowly lowered to begin his examination of the underside of the ship. For a few minutes, Jim watched the bubbles floating to the surface. Finally, just before seven, the speakers broke the silence with the news that there was some slight damage to one of the two propeller blades.

"It will take a few hours to repair the blade and then we will be underway. Thank you for your co-operation during the night. Please make your way to the dining rooms for breakfast."

Icebergs and the propeller dominated the conversations at breakfast and later, when people were strolling on the decks. As the ship bobbed silently in the still sea, Jim found himself talking to a young ensign.

"I guess it was our turn," the young ensign sighed. "Canadian Pacific has three ships just like this. Last July, the *Montcalm* hit a berg in these waters and bent a propeller blade, just like us. Last month, the *Montrose* had extensive damage to its stem and bow when it collided head-on with a 'berg. Fortunately there was no harm done below the waterline, but they lost their anchor in the collision."

"Do you expect we'll see any more icebergs?" asked Jim.

"No, we'll be through the Cabot Strait and into the Gulf of St. Lawrence before dark tonight. They don't float upstream. They come down from Greenland and drift through 'Iceberg Alley', as we call that lane between

Labrador and Newfoundland, and then continue out into the Atlantic where they'll melt over the summer."

Following the tense twelve-hour delay, the ship's crew tested the propeller blade and all seemed well. The *Montclare* slowly picked up speed and headed toward the sun and the fresh water of the St. Lawrence. Before supper, Jim caught his first glimpse of Canada, specifically Nova Scotia — New Scotland. *What a bonnie sight it is.* He had been reading some of the history of Canada in the ship's library and had become quite familiar with the geography of his new homeland. He surmised that the next bit of land was Cape Breton Island, which was often compared to the 'auld sod', being like it in topography and vegetation. As darkness fell on their last night at sea, he could see the outline of the Gaspé Peninsula off the port side.

When Jim awoke the next morning, he found it hard to believe that they were now on a river. *The shores are so far apart. This is a big river. And I thought the Devon was big when I used to sit on the banks and fish with my father! It would not even qualify as a stream here.*

He tucked into his final breakfast on the ship — fresh orange juice, sausages, eggs, toast and jam with real berries in it, pan-fried potatoes, fruit and coffee laced with real cream.

During the morning, the *Montclare* made its way slowly up the narrowing waterway. Shortly after lunch, the ship made a sharp left turn around a large island, and Jim's destination was in sight: Quebec City! As they approached the port, with buildings along the water, a large hill behind them and a kind of castle overlooking it all, Jim thought about how similar this was to Greenock.

"It's the Chateau Frontenac," said a returning Canadian lad whom Jim had befriended. "It's a luxury hotel owned by Canadian Pacific. They have them all across the country. They also own the railroad you'll be travelling on for the next five days."

I don't even want to think about more travel right now. I just want to walk on land for a bit.

The grin on Jim's face as he descended the gangplank was a mixture of delight, anticipation and relief. Ahead, a sign read, '*Port de Québec / Bienve-nue au Canada / Welcome to Canada*'. He was now in Canada, his new home!

Prior to disembarking, the four young men had been told to gather at a designated section of the pier and wait for instructions. Jim waved to some of his partying buddies from cabin class who were being directed through to a different section of the pier. The majority of the passengers who had travelled on the lower decks were herded to an area where a number of interpreters were on hand to help them.

The group Jim was in was smaller than the rest and the English-speaking officials from the Canadian Pacific Railway treated them with respect. After their papers were checked, Jim and Dunc were simply instructed to make their way to the railway station, a number of blocks northwest. Since neither had enough money to pay for transportation to the train station, they each had to hoist their heavy cabin trunk onto a shoulder and set out.

Jim set his trunk down in front of a shop after walking a few blocks.

"Christ, this thing is heavy," Jim grunted. "I need a fag and a chance to get my breath."

"Me, too," said a relieved Dunc.

His own trunk was almost unbearably heavy, but he hadn't wanted Jim thinking he was a pansy.

In time, the two continued the long hot trek, with frequent stops and finally saw the green-roofed railway station building with the sign 'Québec' — their destination. When they went to check in and were informed that there would be a two-hour wait until the train departed, Jim suggested having supper. They soon discovered the station restaurant and wearily sat down for their first meal on Canadian soil.

Jim had just converted his five pounds into Canadian dollars and was aware that it had to last until he got to Edmonton. Someone suggested they order the *special*, as it was usually reliable and the cheapest on the menu. The soup arrived first. Jim loved salt and always sprinkled it liberally on whatever was served to him prior to tasting his food. He reached for the tall glass container with the hole on the top and poured an ample amount of white crystals into his hot bowl. He had earned himself a bit of a reputation as a leader on the trip over so when some of the other diners from the ship saw his action, they followed suit.

"Holy God, this Canadian soup is awfully sweet," Jim gasped as he swallowed the first spoonful.

That was how he learned the difference between a salt shaker and a sugar dispenser — but this experience did not stop his salt habit. His Salisbury steak, smothered in onions, awash with gravy and *properly* salted was closer to what he might have had at home — he wiped the plate clean with his fourth slice of white bread. The custard pudding was an added treat and Jim felt 'sufficiently sufficed', as his father would often say.

When the PA system alerted them that it was now time to board their train, Jim turned to Dunc.

"Come on, lad," he urged. "If this Canadian Pacific train is half as good as the ship, we're in for a swell ride across this country."

After they were seated, the conductor came through the car and handed each person a blanket, a pillow and cutlery, explaining they were gifts from Canadian Pacific to make their trip across Canada more comfortable.

As the train pushed westward, Jim gazed at his reflection superimposed on the luxuriant landscape that was rural Ontario and washed down his crudely created cheese sandwich with a long swig of warm milk straight from the bottle. This was his first breakfast in Canada and he pondered how different things were on that Sunday morning in May 1928, than they had been just twenty-four hours earlier. Licking the white milk mustache from his upper lip, he turned from the window and looked across at his friend.

"Dunc, this is a far cry from yesterday on the boat. This is more like back home."

"Ship, Jim," Dunc corrected him. "We won't get the same kind of pampering here."

The luggage racks that doubled as overhead bunks at night made the long narrow aisle even more cramped. And it was crowded with people, families with young children and a great many single young men. Jim could hear familiar clinking and clanging behind him as the immigrant women prepared meals for their families on the stove at the rear of the car. Since there was no dining car on this crowded train, Dunc and Jim were responsible for their own food for the first time in their lives. They were fortunate that one of the passengers had asked them if they had purchased any supplies for the trip before they left Quebec City, so they at least thought to copy what others were doing and bought bread, cheese and milk. Jim had never cooked anything in his life and had had no need to, with a mother and sisters who doted on him. Now he sorely missed what he had taken so much for granted. So, when lunchtime rolled around, he worked his way to the stove end of the car. There he turned on the charm and discovered that language was no barrier between him and the delicious bowl of soup offered him by a smiling woman in a babushka.

As the journey continued, some of the women in the car offered suggestions as to what the lads could buy with their meagre funds. The train had made many stops as it wound its way across Quebec, through Ontario and then into Manitoba, allowing time for passengers to replenish supplies from the small stores that were within walking distance of the station. The crowd often enabled Jim to bypass the cash register on his way out, as a result he was able to stretch his five pounds fairly well.

"They won't miss a few cans of beans," he offered to Dunc, when they were safely back on the train.

As the car was full, he had to learn patience when waiting his turn at the stove, but he soon mastered the intricate art of heating a can of pork and beans, the result of which brought back memories of his Sunday mornings blasting with Sandy. He did not cook every day, as there often appeared an angel in an apron with something extra for him and his pal.

When they reached Winnipeg, officials from the Department of Agriculture boarded the train to process their immigration forms.

When one of the officials scanned Jim's documents, he said, "You are going to a farm in Stettler."

"No, I'm not. I'm going to Edmonton," replied Jim.

"No, no! You are going to Stettler, young man."

"Look here, they told me in the old country I was going to Edmonton and that's where I'm going. That is where my ticket takes me," Jim said, refusing to give way to the official's pompous insistence.

Looking at the forms again, the man asked, "You're going to a farm?"

"Aye, to a farm at Edmonton."

"Look," sighed the official, "have you ever been to Glasgow?"

"Aye, a number of times."

"Are there any farms in Glasgow?"

"No, it's a city."

"Well, there are no farms in Edmonton. You are going to farm outside of Edmonton."

"How far is Stettler from Edmonton?" Jim asked.

"About ninety miles," responded the official, smiling smugly at catching up this young man.

That's more than from Tillicoultry to Glasgow and back.

"Okay," he said, sensing that this was one official who liked to win. He nodded his head and smiled.

It took twelve hours to complete the form-checking for all the passengers, and when the train whistle finally sounded and the first chugs of the straining engine echoed through the station area, the frustrated immigrants cheered. Slowly, they left Winnipeg and officialdom behind.

They rattled across the prairies, the train stopping briefly at small towns to take on water and allow the passengers to purchase food.

"There aren't many villages," Dunc said.

"But the farms look awfully big," Jim replied. "There should be lots of work for us."

They got off the train in Calgary to stretch their legs and, as they milled about on the platform, they heard an announcement that those heading for farms between Calgary and Stettler would have to change trains.

"What are we going to do, Jim?" asked Dunc, who was also slated for a farm outside of Edmonton.

"Well, the rest of them can do as they please. My ticket is for Edmonton and I am going to stay on this train to Edmonton."

Dunc, though somewhat apprehensive, agreed to follow his friend's lead and stay on the train. Still, not wanting to take chances, prior to boarding Dunc phoned his cousin, Red Morrison, who had emigrated earlier and was now working on a farm south of Edmonton, near Ellerslie. Red assured him that there were many farmers looking for help and that he would borrow his boss' truck and meet them in Edmonton.

The next few hours dragged by for the would-be farmers as the train moved through the rolling land of south-central Alberta. At last, the train slowed as it approached the outskirts of the city. Shortly after they crossed high over the North Saskatchewan River, the lads could see the station ahead. '*Edmonton*', the long-awaited sign read.

"There he is, Jim!" Dunc exclaimed, showing more animation than usual as he pointed to a grinning redheaded man on the platform. "There's Red. Och, it's good to see him."

Red had talked with one of the agricultural representatives while waiting for the train and found out that the two could work together on a farm in Winterburn just a few miles west of Edmonton. He suggested that they check their trunks in the baggage area and find a place to eat. Red steered them down Jasper Avenue to the Pantages Vaudeville building and through a doorway under a sign that said '*American Dairy Lunch*'.

This was another first for Jim, as he had never been in a cafeteria, but he still had 75 cents left and could splurge. What an absolute delight it was to fill his tray from the mouth-watering array before him. Nick Spillios, the Greek owner, grinned as he watched the trio demolish his creations. After Jim swallowed the last scrap, he took out his tobacco and began to roll a cigarette.

"This is the way to live. That was a great meal." He paused to light the fag. "And worth every penny of the forty cents it cost me."

"Food is very important to Jim," Dunc explained to his cousin. "His father always says, 'Starved pups make hungry dogs.'"

When they left the restaurant, Red telephoned the farmer to let him know his new help had arrived. About an hour later, a middle-aged man, wearing black pants held up by large suspenders over a plaid shirt, a pipe clamped firmly in his teeth and a felt hat tilted back on his head, drove up to the station in a small truck.

"My name is Vandenhoff and you will be working on my farm for the next year. Put your luggage in the back."

The two lads gave their names to their new boss, quickly hoisted their trunks up into the truck and climbed into the cab.

"You have seven cows to milk first thing in the morning, harness horses for the day's work, clean out the barn and the chicken coop and collect the eggs," their boss informed them as soon as they were underway.

He went on to explain that there were seven other hired hands and a foreman, whose wife was the cook and looked after the bunkhouse in which all of the hired help lived. Their meals were to be eaten in the foreman's house, starting with breakfast at seven the next morning.

Jim was unusually quiet as he listened to this gruff Dutchman talk of nothing but work. Tired after the long train trip and large meal, he nodded a bit and hoped there would be a comfortable bed awaiting him. There was, but he was not in it nearly long enough as the workday began when the foreman hollered them awake at 5 a.m. and they were led to the cows.

Jim had never milked in his life so Dunc, who had worked briefly on a farm outside of Alva, showed him what to do. He managed the milking but could not strip the last milk from the teats, so Dunc had to do that also. The big pails of warm milk were very appealing to a young empty Scottish stomach waiting for the seven o'clock breakfast. Jim blew the froth back, picked up a pail and drank in the nourishing liquid.

"Boy, was that ever good," he gasped when he came up for air.

They took the full pails from the barn to the coolhouse, where some of the milk was poured into large jugs for the breakfast table and the rest stored to await shipment to the dairy. After the milk in the jugs had settled, the foreman showed Jim how to slowly pour the cream off the top into a quart jar.

"That will be swell for porridge," Jim said.

"It's for the coffee," the foreman replied.

After the foreman left Jim to finish the job, he took a few swigs from the jars and savoured the richness of the warm cream.

When he entered the dining room for breakfast, Jim was startled to see that the cook, a nice-looking young woman, had her dress tucked into her bloomers and a child nursing at her breast while she tossed pancakes. He had never seen anything like that and did not know where to look. His mother had never nursed in front of him or anyone else, and he was certainly distracted by what he saw. Fortunately his stomach won out over his curiosity and he dropped his eyes to the food. There was porridge and boiled eggs and a mound of pancakes on the table. Jim quickly filled a large bowl of porridge and, putting a couple of the eggs on his plate, looked around.

"What do you need, laddie?" asked another of the workers, an immigrant from Aberdeen.

"Where's the bread?"

"Och, you dinnae get bread in the morning. You get bread at lunch and again at supper but you've got to eat hotcakes for breakfast," replied his fellow Scot.

Never one to let food pass him by, Jim decided to try these weird Canadian customs right from the start. He ate seven hotcakes with his seven boiled eggs and felt that it would suffice until lunch.

After breakfast, the foreman directed them to the horse barn.

"Do you know how to harness a horse?"

"No, but I'll learn," Jim replied.

Perhaps he spoke a little too hastily, as the task was not easy for one as short as he was. He could not reach over the back of the large beast and, once again, Jim was faced with being singled out because of his size.

"There's a box. Stand on it and you'll do fine."

Jim reluctantly took the foreman's advice and set about finishing the task as quickly as possible.

It was not until the foreman told him that his next job that morning was to gather eggs that Jim learned that there were about 1,500 chickens that not only laid eggs but would provide him with the substance for his next task, cleaning the chicken coop. It was a dirty job with the fine, stinking dust — more acrid than coal dust — invading his nostrils and lungs. By eleven o'clock, he was hungry, broke open a few of the fresh eggs, swallowed them down whole and smacked his lips. *That'll do for now!*

The working, sleeping and eating pattern continued for about three weeks. Then one night after supper as Jim and Dunc were sitting down to write letters home, the foreman came in.

"The boss wants to see you boys now up at the big house," he said.

"What do you think, Dunc?" Jim asked, as they walked toward the owner's home. "Are we getting the boot?"

"I don't know, Jim. You have been drinking a lot of cream and eating more eggs than ten hens lay each day."

"Och! Dinnae be daft, man. You're the only one who knows about that and you haven't been talking, have you?"

"No! You know I wouldn't do that."

Vandenhoff and his wife were outside waiting for them. Jim felt more comfortable when she was around as she was from Hamilton in Scotland and her speech was always a comfort to him. The boss' voice evoked the opposite feelings in Jim.

"It's time to put the garden in," Vandenhoff said gruffly. "It's all ploughed and ready for you."

He then laid out the details of how he wanted it all done. Jim and Dunc got at it immediately and worked until dark. More tired than usual when they returned to the bunkhouse, they washed up, crawled into bed and slept soundly until five the next morning. The next evening as Jim settled to write his letter, the foreman arrived again.

"Come on, lads, the boss wants you up at the house."

When they arrived, they weren't greeted fondly.

"What's the matter with you?" Vandenhoff exclaimed. "I expected you here right after supper. There's work to be done in the wife's garden."

"I didn't know we were supposed to do this as well as the other work we do all day," offered Jim.

"That's your job," barked Vandenhoff. "It's part of the deal, the contract I have with the government. You get $30 dollars a month with an extra dollar a day during harvest, and you do whatever I decide is necessary around here," the boss barked.

"Do we get paid extra for this garden work?" asked Jim.

"That's part of your $30," he replied somewhat impatiently.

Dunc in his usual quiet manner said nothing during this exchange, leaving it all to Jim.

"To hell with this!" Jim shouted. "I'm up at five in the morning and it's seven at night when I finish supper. I'm not doing this for nothing extra!"

"You can't get out of it. You said you'd work a year and that means doing whatever I tell you to do."

A conversation with Red Morrison popped into Jim's head. Red had told Jim and Dunc about a Major Jack Miller, who was a land officer with the CPR. He told the boss firmly that he was not going to work that night and that he was going to Edmonton the next morning to talk with Major Miller.

"Do you know Major Miller?" asked the boss.

"He comes from my hometown in Scotland," Jim lied.

He had never met Major Miller or even heard of him until Red had talked about the man.

"I'm through. I'm not taking this anymore," he asserted with some finality.

The argument continued for some time but Jim would not back down, and finally the boss agreed to give him his pay in the morning. Jim headed off to the bunkhouse with Dunc beside him.

"What have you done, Jim?" he asked. "We are going to be in real trouble with the Canadian authorities."

"Don't worry about it, Dunc. When they hear our story, we'll be fine."

The next morning following breakfast, the foreman told them to put their trunks into the truck as he was going to Edmonton to deliver eggs. When they had done this, they went up to the big house for their cheques.

"What the hell's this?" Jim asked as he looked at his cheque made out for $15.

"Well, you worked part of the month and I've figured out what I owe you."

Jim told to him that they had worked more than half a month and they should get more money. Another argument ensued. The reluctant boss finally agreed to write new cheques for $18.75. This was not as much as Jim wanted, but somehow it felt like a victory over "the old bugger".

There was no work immediately available for them in Edmonton, so the two young men found accommodation in a rooming house. One of the roomers suggested they try the 'slave market', a place where casual labourers were hired each day. After a couple of days doing odd jobs, they heard that a Norwegian named Johnson was looking for help on his farm. This was about ten miles east of Leduc, which is about thirty miles south of Edmonton, and so was in the area where Red Morrison was working. Jim did not hesitate to take the job and Dunc, called Red, who told him of another opening on a farm about three miles from where Jim would be. All three decided it would be a real bonus to live closer together.

Jim asked for the same remuneration as he had previously been promised, $30 a month with an extra dollar a day during harvest, and his new boss agreed. Johnson owned a small sawmill a few miles from the farm and told Jim he could make extra money by cutting firewood. Jim soon learned how to put up fences and drive a team of horses pulling discs and harrows.

Mrs. Johnson was a lovely woman with two young boys at home, and she invited Jim to attend the community church with the family. Jim met Dunc and Red there each Sunday, and following church, they often walked to the farm where Dunc was working and would have dinner with the family there. The food, served by the English people Dunc worked for, was more familiar to Jim than the Norwegian food of the Johnsons, so it was well worth all the walking he did each Sunday. Red hitched rides back and forth each week, as he lived about fifteen miles farther north.

Johnson also owned a homestead east of his place in New Sarepta. One hot summer day, he decided that he and Jim would go out and break new ground in a field he had recently logged. They loaded a wagon with food and supplies, took four horses and set off. The only building on the site was a shed where they could cook and sleep. Mr. Johnson broke ground with a team of horses pulling a sod-busting plough. Jim followed with another horse hooked up to a stone boat and picked up rocks, tree roots and other debris that would get in the way of a growing crop. It was hard work loading and unloading the stone boat all day and the hot sun drained his energy quickly. He could only keep at it if his stomach was full, but there was no way to preserve their food. By the second day, the pork they had brought was getting a bit soft, and they had little else to eat but porridge and coffee. After a few days of long hours, poor food, hot sun and a bed of hay next to the horses, Jim began to rethink his situation.

Things came to a head on the day Johnson came upon a huge stump while he was ploughing. He left Jim with a pick and an axe and told him to dig and chop up the roots, so they could remove the stump with the team. Meanwhile Johnson ploughed another round of the field and, as he passed by Jim, asked how he was doing.

"It's a tough one. It won't come easy."

"Well, keep at it," replied Johnson.

On the next round, he stopped again.

"Goddamn! Have you not got that root out yet?"

"Do you think you could take it out any quicker?" Jim snapped.

"I guess I could," he snarled back.

"Well, I'll be goddamned if you don't. Here's the axe. You go to it!"

Reaching an impasse in this heated discussion, the two men softened a bit. Johnson went off on another round and Jim attacked the tree roots once more.

As Johnson completed another circuit, he made another sarcastic remark as to Jim's progress. Jim was boiling with the heat, hard work and his boss' attitude, and although he had the roots almost clear, he jumped out of the hole he had dug around the tree.

"Here. Take your bloody axe and your pick and do it yourself."

Johnson paused for a moment, looking at the dripping young man.

"Simmer down," he said. "It's too hot to work anyway. Let's go in."

As Jim talked about the work and the grub, Johnson agreed to return to his home about nine miles away and get fresh food. He rode off on one of the horses while Jim cooled his heels and other simmering parts of his body in the shade.

That night, just as it was getting dark, a thunderstorm struck with fury and added the punch of hailstones to the flashing lightning, roaring thunder and driving rain. Jim had never seen the like before, and when a bolt of lightning hit a nearby tree, this brash young immigrant became truly frightened and felt intensely alone. He huddled in the shelter and buried himself deeply in the hay. It was one of the longest nights of his life, and he was overjoyed when he saw the horse and rider approaching early the next morning.

Johnson arrived with fresh eggs, bacon, bread and more coffee. It was a great meal to start the day, but as they continued to break ground, they encountered another big tree. The pattern of the previous day began again, with Jim working as hard as he could in the heat and Johnson adding fuel to the burning temper of the young man with his sarcastic interjections, delivered sharply on each round of the field.

Jim had had enough.

"Look, why don't you stop your bellyaching and unhitch those horses and pull out the bloody tree?"

"I can't take time for that. I have to get the breaking done," Johnson responded.

"To hell with you. I'm going."

With that, Jim began walking back to the shelter. Johnson followed.

"Don't quit. I guess I am a bit hard. Must be the heat," he pleaded, trying to coax him back.

"You're too tough a guy to work for," Jim was adamant, "and I won't stay another day!"

No amount of pleading was going to change this determined young man's mind and Johnson agreed to leave the project. They collected all of the stuff they had brought out, loaded the wagon and drove silently back to the farm.

"I'll leave in the morning," Jim said crisply when they arrived.

Following breakfast the next morning, Johnson gave Jim his money — the correct amount.

"Are you going to give me a ride into Leduc?" Jim asked.

"Damn. No! If you quit, you can find your own way," Johnson stated firmly.

"I've got a big trunk." Jim pleaded.

Even with his wife taking Jim's side, Johnson was adamant.

"You find your own way."

Jim set off with his trunk on his shoulder and headed down the road a mile and a half to the baseline road. It was a lot warmer than it had been in Quebec City, and he had to pause a number of times before he reached the crossroads. He was sitting on his trunk, exhausted, trying to catch his breath, when a little truck came along with a nice-looking girl at the wheel. She stopped and asked Jim where he was going.

"Leduc," he replied.

"So am I. I have cream to take to the dairy there. Hop in."

Jim found renewed energy with the young woman watching, and lifted his heavy trunk up into the back beside the cream cans. The short but pleasant ride eased some of his weariness, and when he was dropped off at the train station, his disposition was much improved.

He caught the CPR train to Edmonton and once more disembarked at the main station on 109th Street. From there, he walked along Jasper Avenue with his trunk on his back to 101st Street, where he wearily lowered the trunk and sat on it long enough to roll and smoke a cigarette. With a loud grunt, he once again shouldered the trunk and continued his search. Two blocks later he spied a house with a 'Room To Let' sign in the window. He quickly arranged a room and, dog-tired, he settled in for the night. The next morning, he took his trunk to a second-hand store just around the corner on Jasper Avenue and exchanged it for a kit bag large enough for his things. He was no longer going to travel around this new land with a steamer trunk.

One of the roomers told him that it would be hard to find work because there were too many immigrants from Europe arriving every day. However, Jim had no choice but to find a job as quickly as possible. The next morning, he came upon a construction project on the corner of 101st Street and Jasper Avenue, where they were preparing the foundation for the new Royal Bank. The workers used picks and shovels to excavate the hole, the debris was hoisted to the surface by a manual pulley system, and then others moved it by wheelbarrow to the trucks that hauled it away.

With so many unemployed men hanging around the site, it was obvious that, once hired, a man would have to keep working steadily, as there was always someone waiting to take over. He spent the whole day waiting for an opportunity to get some work. Nothing happened. The next day he got up early and finally got to the head of the line at eight o'clock, he met the foreman, a big Englishman.

"How's chances for me?" Jim asked.

"You're not a bohunk?" the foreman said with a puzzled look.

"No, I'm sure not."

"Well, you're too small, anyway. This is heavy work and we need people with strong backs."

"Just give me a chance. I've been in the mines since I was thirteen," Jim replied, looking up at the big man.

"Okay, grab a wheelbarrow and we'll see what you can do."

That was all Jim needed to hear. He had watched the pattern and knew that as long as you had a wheelbarrow and had been signed on with the timekeeper, you were on your way.

He began the routine: fill the wheelbarrow, go as quickly as you can to where the contents were to be dumped, run back and do it all over again. About ten o'clock, he began to drag a little. A number of men had already dropped out from the strain of the hard work but Jim pushed on, knowing that he could not show any signs of slackening.

"Hey, Jock, come here!" hollered the foreman.

Jim held on to his wheelbarrow as he responded to the command.

"Drop that wheelbarrow!"

The foreman then called over a much larger man and told him to take the wheelbarrow and get to work.

"What the hell's the matter?" Jim feared he was about to be fired. "Aren't I doing enough?"

"I've got another job for you."

"What's that?"

"Go down inside those forms on the other side and pick up all the scrap wood that's scattered around. It has to come out before we can pour the concrete."

Showing his eagerness, Jim scampered down to the bottom of the construction site. The first thing he noticed was how cool it was compared to the surface, a bit closer to the kind of working conditions so familiar to him. Thirty minutes later, he climbed back up to the surface to get a cigarette from his jacket.

"What the hell are you doing up here?" the foreman bellowed.

"I want a fag."

"I don't know what the hell you're saying, Jock, but get back into that hole and get to work."

Jim took his jacket, returned to the bottom and lit his cigarette in the coolness of the morning.

The job lasted about five days and Jim did well. At twenty-five cents an hour, it was far more than he had earned on the farms and the hours were better. He was even able to save a little. However, when that job finished, he had to look for other work.

Even though twenty-five cents could buy him a meal of liver and onions, bread, potatoes, gravy and coffee, his meagre savings were soon depleted as each workless day followed another. Soon he had to buy fifteen-cent meals and then came the day when his money was finished. He was almost broke with only a couple of dollars to keep him in cigarettes for a while, but he had to move out of the rooming house and find a spot to sleep in the brush beside the river. When he had spent the last of his funds to buy a loaf of bread and a chunk of bologna, he became very discouraged, homesick and ready to give up.

Alone and desolate, wearing his old country suit and cap, he stood at the corner of 101st and Jasper one morning, not knowing what his next step might be. *Canada is not the glorious place I thought it was.* He was jolted from his despair by the greeting of a big man who had stopped in front of him.

"Hello," said the stranger.

"Uh . . . hello."

"You're Scottish, aren't you?"

"Yah, what's it to you?" Jim said sullenly.

Even though he recognized the voice of a fellow Scot, he was in no mood for idle conversation. Suddenly his stomach loudly reminded him that it was being neglected again.

"Are you hungry?" the man asked in response to the rumblings.

"I sure am."

"Well, come with me and I'll get you a meal," the man offered.

Jim remembered his mother's admonition about getting involved with strange men. *But I'm hungry and I've got a good pair of boots on my feet.* He looked up at the big man.

"I could sure use some good grub," he said.

As they walked toward the man's rooming house, Jim learned that he was a CNR policeman by the name of Don Henderson. The landlady invited

Jim to join the others for lunch and he gratefully devoured the much-welcomed meal, another of those memorable hungry-dog experiences.

When they had eaten, Don said, "Well, I have to go back to work. You go and lie down on my bed and have a sleep while I'm gone."

Jim quickly fell asleep in the first bed he had seen in over a week. When Don returned about six o'clock that evening, he had to wake the exhausted young man.

"Come on, it's time for supper."

Jim quickly washed his face, gratefully downed another full meal, and then spent the rest of the evening in conversation with Don and others in the rooming house.

"I'd like you to come round to the station in the morning." Don said before he went to bed. "I'll take you to the hiring office to see if there are any jobs."

Jim slept on the floor that night and, following a hearty breakfast, headed out with Don. The timing was perfect, as the company was in the process of forming a new extra gang to build a section of railway east of Edmonton on the line to Lloydminster.

"Have you ever worked in a kitchen?" the official asked Jim.

"No, I haven't. I've never worked for a railway either."

"Well, you are going to be a cookey."

"What's a cookey?"

"You'll wash the dishes and serve the tables for about 150 men on the extra gang."

"That suits me," said Jim. *If I'm in the kitchen, at least there will be plenty of food.*

"You'll have to get a blanket and a couple of white aprons."

"That's fine. I'll do that," Jim smiled.

When he came out of the office, Don Henderson was waiting for him.

"Well, did you get a job?"

"Yes, I did. I need a blanket and some white aprons though, but I have no money."

"Here," Don said, handing Jim two dollars. "You can get what you need in the store down the street."

With his new purchases stashed safely in his kit bag, Jim got on the work train and found the kitchen car. It was hooked up between the one that the cooking crew slept in and the large diner in which everyone ate. The work gang was mostly central Europeans who spoke little English and, in Jim's eyes, looked very big and mean.

Dreams Unfulfilled

"Get your apron," the cook said.

He prepared to make mounds of sandwiches, as Jim went to his bag and put on the white garment.

"What the hell are you doing in that thing?" the cook demanded as he stared at Jim in his woman's bib apron. "Give that to me and take a couple of mine."

The cook put the offending item in his own bag to take home to his wife. About halfway out to the site, the train had to stop for the lunch break because there were no connecting doors to the crew cars. Everyone disembarked and lined up outside of the dining car for sandwiches.

Jim worked consistently, helping the cook, washing dishes and serving food. He ate at least his share and savoured every morsel he ingested, but he lost his job after a month of hard work when a man with more seniority bumped him.

"This is a hell of a way to run a railway," he snarled at the cook when he received the news.

"It's the way we do it, lad. Keep in touch at the station. Something else may turn up."

Jim had no way to get back to the railway yard in Edmonton and was forced to experience another first — riding the rails. Following instructions from one of the work crew, he ran alongside a freight train as it moved slowly past the work crew. He found a boxcar with a door open, threw his bag into the car and then leapt up and scrambled onto the floor of the car. He jumped to his feet and leaned out of the door to wave goodbye to his grinning tutor.

In Edmonton he picked up his $45 cheque for the month's work and immediately found a place to stay in a rooming house. It was the August of his first summer in Canada and he did not want to sink as low as he had done previously, so once more he began to walk the city streets each day, looking for anything that would change his course. As the temperature rose under the hot summer sun, his spirits diminished along with his savings.

One day as he was walking along Jasper Avenue, he encountered an older man — another Scot named Willie — who recognized the lad as an immigrant. As they talked, Jim told the man how hard it was to get steady work and of the variety of experiences he'd had in his first three months in Canada. Willie told him that he was the blacksmith at a mine in Clover Bar, just east of Edmonton.

"Why don't you get a job in the mines?" he asked.

"I left Scotland to get out of the mines," Jim quickly shot back. "There's gotta be something else to do."

"Look, lad, I've seen a lot of young men come and go here and there's just too many looking for jobs."

Jim stood silently for a few moments. He had experienced hunger, hard work in the heat, and the uncertainty of the next job. He was living from hand to mouth and had no prospects of changing that. He knew that, if he had to, he could work in the mines until something better came along.

"Where's this mine you're talking about?" he inquired.

Willie told him, but emphasized that it was a bit early in the season for the mines to be hiring.

"Where are you from, lad?"

"Tillicoultry."

"Och, do you know the Dunn family in Sauchie?" asked the old Scot.

"Not really," replied Jim. "Though I know of a minister by the name of Dunn there."

"Well, that's the same family," he said. "The Black Diamond mine in Clover Bar is owned by the Dunns. It's part of the Alloa Coal Company."

"I worked for them in Dollar!" said Jim. "I have a letter of reference from the manager there."

"Look, why don't you go and see Alex Dunn, the manager?' he suggested. "He's in the Black Diamond office on 102nd Avenue."

The next morning, Jim was shown into Dunn's office.

"I'm looking for work," he immediately declared. "Do you have any jobs at your mine?"

"No. Not for another month yet," Dunn replied quickly.

"They told me you might have a job. I have a reference from the Alloa Coal Company in Dollar."

Dunn took the paper.

"Oh," he said, "maybe we can find you something. Go out to the mine and talk with Jock Davidson, the pit boss."

Early the next morning, Jim walked east to the edge of the city, through Beverly, carefully negotiated the high railway bridge over the North Saskatchewan River and then cut across country to Clover Bar. It took four hours of brisk walking to reach the Black Diamond mine.

"I'm sorry. There are no jobs," Jock told him after he had explained his reason for being there.

"Didn't Alex Dunn tell you I was coming?"

"Oh, so you're the guy. I thought you would be . . . uh . . . older."

Dreams Unfulfilled   141

"I'm twenty and I've been in the mines almost seven years," Jim replied, once more stretching to his fullest height.

"I might have a job for you. Have you ever laid track?'

"No," Jim hesitated. "I was always at the coal face."

"Well, I'll put you with my best track layer and he'll teach you."

"I'd really like to get on the coal. That's where I can make the most money."

"Well, there is no digging right now. You'll just have to take what I'm offering you. Once the orders start coming in for the winter, there'll be more chances. This is all there is for now."

"I'll take it," said Jim. "I'll learn something new while I'm waiting for a job at the face."

Davidson was impressed with this eager, almost cocky, young man.

"Do you have a place to stay?"

"Not out here. I have a room in Edmonton and it took four hours to walk here today."

"You say you come from Tillicoultry?"

"Aye."

"Do you know Granny Hunter?"

"Sure." responded Jim. He pictured the old woman who had often visited with his mother before leaving for Canada to be closer to her son.

"She has a place near here. Go over and see if she has room for you."

Granny, who lived in a tiny one-bedroom house, was delighted to see the son of her friend and welcomed him in. She would have kept him all day, as she relished the news of her hometown, but Jim, anxious to get his digs squared away, asked if she knew of a place for him to stay. She apologized for not having enough space for him but told him of another Scottish family, Sam and Effie Davidson, who might have room for a lodger.

"Sam lost a leg in a mine accident so he's now the night watchman at the tipple," she explained. "He's not making as much money as he used to. They have a couple of miners boarding there."

Jim promised to come back and visit another time and hastened to the Davidsons. He was invited in to share lunch and talk over the terms of living in their home as a boarder. The long walk back to Edmonton was much easier than his trek earlier that day. The following morning, he hoisted his pack on his back and returned to his new home in Clover Bar.

When Jim started work two days later, he experienced another first — the mine had a washhouse. He had never had a shower and could not believe how luxurious it was to have hot water pouring over his head and running down his body as he rubbed Lifebuoy soap into every inch of his skin — except his back. Not only did he feel clean, but he smelled almost antiseptic with the perfume of the germ-killing soap filling the shower room and following him as he put on clean dry clothes for the short walk home. How different this was from the three-mile hike in the cold back in Tillicoultry, with wet clothing clinging to his freezing body. If he had to be in the mines in Canada, there were some things that made it a little easier to take.

Dunc, who was still working on the farm near Leduc, kept in touch with Jim by mail and asked him to look out for a job for him nearer Edmonton. Within a month Jim found his friend a job on the tipple. Dunc had never worked underground and was delighted to have an opportunity for higher pay and accessibility to the city. The Davidsons agreed to have him stay at their place, increasing the number of boarders to four. Effie treated the boys as a mother might. She cooked, washed their clothes and looked after their banking, taking what she needed for board, then, after putting some into a savings account, giving them the rest to spend.

As the mine operated all winter, Jim worked steadily and was able to increase his savings while sending his mother $25 each month, increasing it to $50 at Christmas. Early in the new year, a letter from his brother Jock advised him that their mother was worried about the source of the windfall and wondered if Jim might have broken into a bank. Jim's next letter contained more detail of the work he was doing and the pay scale in Canada.

Meanwhile, Mary Wilson's sister, Annie, had emigrated and was now living in Edmonton. Jim was surprised to see her when they met at a winter dance in Clover Bar.

"What do you hear of Mary?" he asked following their initial greetings.

"She's well and always asks if I've seen you. Don't you write?"

"Once in a while . . . It's enough just to keep up with my mother. Have you convinced her to come?"

"With me away," Annie said," she feels she has to stay with mother."

"Well, I've decided to stay in Canada. Do you think she might come if I went for her?"

"Och, I don't know, she is quite a homebody. She misses you."

But not enough to move. Jim, feeling some resentment rise up in him, changed the subject.

Chapter Eleven
Riding the Rails

In March 1929, the mine shut down until fall. Jim decided that he did not want to touch his savings, so he and Dunc began to look for other work over the summer. There still wasn't much available in Edmonton at that time. As they walked the streets looking for any opportunity, Dunc ran into Major Jack Miller at the CPR station and told him that he and a friend were looking for work. Miller sent them off to see a colleague, Mr. Ritchie, the Deputy Minister of Highways for the United Farmers government, which had been in power since they had defeated the Liberals in the 1921 provincial election. As a favour to Jack Miller, Ritchie decided to give them jobs driving trucks on a highway project east of Edmonton, beyond Cooking Lake.

"Do you have a truck licence?"

"No," they replied in unison.

"You will need to get a licence to do this job."

"I've never had a licence in my life," Jim replied. "What do we need to do?"

"Well, you go to the Highway Department. I will let them know you are coming," offered Ritchie.

The man at the licence office filled in all the necessary forms with the correct answers. In no time, Jim and Dunc got their licences and each had a job.

Jim was responsible for keeping the grader operators supplied with gravel, as they maintained the highway between Edmonton and Tofield, filling in potholes after rainstorms. He arranged a room in south Edmonton for weekends, near the shop where he was to deliver the truck every Friday night for maintenance. He bought a tent, a cot and a tiny cookstove. His plan was for him and the helper assigned to him to camp at Cooking Lake during the week.

His first day on the job, when he went to the shop to get his truck, it was lined up with about twenty others. They were waiting to be backed out of the yard and driven away, but neither he nor his helper knew how to drive. Jim picked up the keys to number 10. *How the hell am I going to get that truck out of there?* Luckily, Jim ran into Alex Simon, a man he had laid track with in the mine, who also worked for the government in the summer when the mines were down.

"Can you drive a truck, Alex?" Jim asked.

"Sure. Why?"

"Where's the toilet?"

Alex pointed out where it was.

"I'm bursting to go. Could you drive number 10 out on to the road for me? They want it out of the way," Jim asked with some urgency.

"Sure."

Jim headed off to the toilet and sat there smoking a couple of cigarettes to give Alex ample time to complete the task.

"Your truck's out there ready to go," Alex said when Jim returned.

"Thanks," replied the relieved young Scot.

He got into the truck, started the engine, and managed to get it into low gear and slowly moved away from the maintenance yard. He drove all the way to Cooking Lake, about thirty miles, in low gear. They unloaded their belongings and set up the tent. Jim spent the rest of the day practicing. He shifted the gears, letting out the clutch as he depressed the gas pedal, then taking his foot off the gas pedal and jamming the clutch down while pushing the gearshift lever into a new spot. The gearbox protested his every move with horrific gratings that sounded as though everything inside it was fighting with everything else. He went around the campsite until he was able to shift in relative silence and then, turning off the engine, he proudly strode over to his bewildered helper.

"Well, man," he announced, "you are now working with a real truck driver."

He stayed on the job until late August, when he heard that the mines were starting up again. Jim contacted Jock Davidson who told him he could come to work right away. Mr. Ritchie was not pleased when Jim said he was leaving the highway work as he wanted him to stay until freeze-up.

"I can't do that," Jim explained. "You know this was only intended to be a summer job, and another month or two on the highway won't get me through. I need to work in the winter. That's where the money is, so I have to quit now, as I have a job waiting."

There was nothing Ritchie could do to dissuade this resolute young man. So he shook his hand, offered a half-hearted thank-you and wished him well. Dunc stayed on while Jim headed off to another season in the mine.

Jock Davidson was promoted to manager that winter. Soon, Jim was rewarded for his hard work and determination by getting a contract to operate the coal-cutting machine. Since earnings were based on coal output and he was a hard, conscientious worker, he had the opportunity to make a great deal of money that winter. Living right next to the mine, he was often called in to work at the last minute when someone did not show for a company job. These extra shifts and the resulting increase in income helped his savings grow even more, and, once again, he was able to send his mother $25 each month.

Jim's social life was similar to what it had been in Scotland, except that instead of the young folk climbing a glen for their parties, they descended to the Saskatchewan River valley below the bridge for some outings. On Saturday nights, he usually attended the dance in the Clover Bar community hall and sometimes took a break from the dancing to join the make-up band, singing a few Scottish songs or playing his harmonica.

Dunc had started dating a young woman named Annabelle, who had a sister and, while they often spent time together as a foursome, Jim did not get romantically involved. He occasionally wrote to Mary and, with his strong sense of loyalty, thought he should return to Scotland one day and pursue the idea of marriage with her. *But if she doesn't want to leave her mother, there'll be no marriage*, he worried. Dunc liked the idea of returning home and agreed that they should try to save enough for a trip the next year. He was now working for the city of Edmonton's maintenance department, where he wasn't making as much money as Jim.

The next spring when the mine shut down, Jim returned to the highways department looking for work. Mr. Ritchie knew he was a hard worker but had been disappointed in his early departure the previous fall, so he offered him a job working on a labour crew building a new piece of highway just a short distance south and east of Clover Bar. Not wanting to spend his savings over the summer, Jim took the job, once again enduring hard work in the Alberta heat, but sticking with it until the cycle brought him back to fall and the mine.

Jock greeted him on his return to the mine in August.

"Jim, I have a new job," Jock told him. "The company wants me to manage the mine in Drinnan. I'll need some good workers I can trust, so I would like you to come along."

"Where's Drinnan?"

"On the rail line between Edmonton and Jasper, right next to a town called Hinton. It's a new mine with high-grade coal that the railway wants, so there will be plenty of work."

"Can I get on the machine?"

"That's why I want you to come. You can out-cut most of the men I've seen and besides you speak the right language."

A few days later, Jim said goodbye to a tearful Effie, and with all of his possessions stuffed into his duffel bag, he boarded the train that would take him to his new home. Jock, who had arrived a few days earlier, drove out to meet the train in Hinton in one of the two automobiles in Drinnan. After quick hellos, Jock drove Jim back to the single men's bunkhouse. It was not as cozy as Effie's home had been, but Jim quickly found a ready audience in the other young miners for his entertaining stories about life in Scotland and why there were only two kinds of people in the world, "the Scots and those who wished they were."

Once more, Jim quickly made his way into the life of the community. He worked hard and steadily, continuing to add to his bank account. During the winter of 1931–32, Dunc and Red met him in Jasper for a weekend holiday and in the course of their conversations, he and Dunc agreed to use their savings for a trip that would take them by train to Toronto, where they could get temporary jobs to earn enough for the return passage to Scotland.

When the mine closed for the summer at the beginning of March 1932, Jim and Dunc set out. When they reached Toronto, they got a room in a hotel and began to experience life in the big city. There were many things to see and do, so, before long, they realized that their savings were dwindling. In the spring of 1932 there was not much work available anywhere, but always enterprising, they put ads in the paper. The only people who responded were farmers and Jim had enough of that. He never wanted to go back on a farm again.

One day, they saw an ad in the paper from the Lipton's Tea Company, offering jobs in sales. What a rosy picture the representative painted of the company, its product, how easy it was to sell tea. It was a glorious opportunity to make big money. They paid the $10 for the steel suitcase filled with tea samples and set out to make their fortune. The generous commission promised them was a great incentive to send them on their way. They travelled around the city by streetcar, but after a day of hard pavement pounding and loud door knocking they had only sold about $15 worth of tea. The next day, they sold even less.

"To heck with this, man," Jim said to Dunc. "We are spending money for nothing and wearing out our shoes to boot."

So they kept the tea for their own use, sold the suitcase, and looked for another venture.

It took only six weeks of a pleasurable lifestyle to spend almost all that they had saved. When they assessed their resources, they discovered that there was not enough money left for the train fare back to Edmonton — let alone the trip to Scotland and back.

"I guess that does it for going home, eh?" Jim was dejected.

"Aye, we'll have to go back and start all over again," Dunc replied.

"I'm not sure I really wanted to go anyway," Jim continued. "I don't think Mary and I are ever going to get married."

"Well, I'm lucky, I guess. Annabelle said she would wait for me to return."

"Maybe I need to find a nice Canadian lass," Jim said.

So they shipped their bags back to Edmonton and hopped a freight train. There were many empty coal cars and they had no difficulty in leaping into one as the train slowly picked up speed on the outskirts of a populated area. The two young men rationed their money so they could travel for a few days, get off, get a bed for the night, wash up, eat a restaurant meal and get back on a freight the next day.

The railway police were diligent in the yards but did not venture too far down the tracks. As they entered Regina, Dunc leapt from the boxcar, but Jim waited a little too long as the train entered the rail yards, and was arrested by a policeman who took him to the local jail.

Each day for a week, the jail keeper led Jim to his house, where he had him paint the fence around his property. The man's wife fed Jim lunch and supper and he quite enjoyed his week in the country with some of the best meals he was to eat on that trip. The day after the painting job was finished, the jailer escorted Jim to the railway yards and waited until a westbound freight began to slowly move out on its journey.

"I don't see you now," he said to his captive.

Jim, assuming that Dunc had managed to get in and out of Regina safely, found a boxcar with the door open and leapt in. Two days later, he was in Edmonton.

Chapter Twelve
Independence

With her schooling now ended, Ida helped out at home for a few months, transferring her anger at her father to the washing. She vigorously scrubbed the clothes by hand, rinsed them and strangled the water from them before hanging them out to dry. Angelina continued to do all of the cooking, finding it easier to do it herself rather than try to instruct her obstinate *I-will-not-speak-Italian* daughter in her halting English.

Frank heard of a family in Sterco, about ten miles along the rail line, who were looking for someone to help care for a little boy and do housework during the final months of the mother's pregnancy with her second child. Ida agreed to go and, although it was her first time away from home, she looked forward to a house with fewer people — and hopefully less stress. She was given room to herself, the first time that had happened in her life. She found that her experience at home with her younger siblings made it easy to look after one small boy and help with the household chores. It was also her first opportunity to have money to spend on herself, and she took advantage of the Eaton's catalogue to buy a few trinkets and clothes with her $10 monthly salary.

Just before the baby arrived, the Roberge family from Coal Valley, farther along the branch line, visited in the home of Ida's employers and asked Ida what she was going to do when her time was finished in Sterco.

"I guess I'll go back home."

"Would you like to come and live with us?" asked Mr. Roberge. "We have three children and my wife is not well. We need someone to help look after the children and do housework, much like you are doing here."

Ida looked at the children, who did not appear to notice how pale and sickly their mother looked.

"I'll have to ask my father, but I think it will be all right," Ida replied.

Her father wrote back, giving his approval. Soon, Ida moved to Coal Valley and began her new job. Mrs. Roberge had tuberculosis and, as the illness progressed, she was able to do less and less, so Ida was expected to do more and more. It was not as easy as the previous situation — three children meant more than three times the work. She did her job without complaint, although in her letters to Stella, she wrote of the family's demands on her time.

Being too far from home to visit, these letters were her only contact with her family. She could not tell Stella much about her social life, as she knew her father would read the letters. Although she had little time off, she did manage to spend some early evenings learning to play tennis.

In the previous coal-mining towns Ida had lived in, almost 50 percent of the miners were of British origin, with the rest Italians, Serbians and Ukrainians and small clusters of Americans, Romanians and Swedes. Coal Valley was an exception to this pattern, where the directorship successfully maintained a stable ethnic composition by insisting that 75 percent of the miners be French-Canadian or Roman Catholic if they were not French-Canadian.

As in most mining towns, the company had built a tennis court and that became the centre of activity during the summer. Ida, being one of the few young women in the camp, was very popular with the young French-Canadians. She learned from the young men that, 'If you have a Slazenger, you really have something.' The Slazenger Company had been established in England in 1881, four years after the All England Lawn Tennis and Croquet Club had formed and held its first matches at Wimbledon. The company began producing 'The New Game of Lawn Tennis' in a box, with racquets, balls and a net. So, Ida sent away for her first racquet, a Slazenger, which cost five dollars.

With the help of numerous enthusiastic volunteer coaches, Ida was soon able to hold her own on the court, enjoying the freedom the distance from her father afforded. How she wished she could share her popularity with Stella.

Then one day, a letter arrived for Ida. As soon as she saw the beautiful handwriting, she knew it was from her father. The news was not surprising. The family was on the move again, this time to a new mine in Drinnan, just outside of Hinton. Mary and Carl would also be moving. Drinnan was a long way from Coal Valley, with the only way to get there being by rail north to Edson and then west again toward the mountains. Stella later wrote that their father had sold his gold chain, which helped him buy some land about two miles from the town between Drinnan and Hinton, and they were building a house. Stella and Johnny only had one mile to walk to school.

Ida's anger rose as she thought of her father making decisions again without concern for anyone but himself. *He always keeps the family away from the rest of the town. It is so hard on Mama. I will never ever marry an Italian!*

One day, six months later, Mr. Roberge came home with the mail and called for Ida.

"You have another letter from your father," he said. "I recognize his handwriting."

"He is very proud of that," replied Ida.

Frank had learned to write in Italy and had to teach himself to write English when he came to Canada, she knew. He still read westerns to learn the language. *Westerns are so much easier on the family. When he reads the newspapers, he gets so mad at what is happening back in the old country.*

As she took the letter to her room, she thought about how peaceful her life was without her father's constant anger.

She began to read. The letter told her — did not ask her — to come home. The beautiful script did not ease the anxiety that began to burble up inside.

I have a job waiting for you working for the mine manager, doing housework and looking after his children. You will be able to stay at home, as it is only about a mile or so from where they live to where we live. I will expect you to arrange for the train and be home within a few weeks.

She read that section again and wondered what to do. Stella had described Drinnan as a bigger town than Mile 40 and said there were many young people there.

A letter from Stella arrived two days later, reminding Ida that she had not yet seen Mary's two daughters, Margaret and Jean, who had been born after the move to Drinnan.

They are delightful babies. I am looking forward to you coming home again.

After weighing the consequences of what might happen if she refused, Ida decided to accede to her father's bidding. With mixed feelings, she told the Roberges that she had to leave.

Once again, Ida was living with her family and had to walk two miles — not the one mile her father had written about — to work each morning six days a week in order to start at 7 a.m. and then walk wearily home when she was finished at 7 p.m. Working for mine manager Jock Davidson was more work than she had before as he, his wife and two children lived in a many-roomed two-storey house with lots of furniture and fancy wood trim to dust.

Still, Ida did well at the Davidsons and they, in turn, treated her with respect and appreciation for her efforts, paying her $15 a month and giving her time off whenever she had a ball game. Soon, Ida was able to buy new clothes for herself and Stella. The next summer, the Davidsons took her with them on an extended trip to Vancouver so she could care for the children and do the cleaning in their rented cottage, but she still had time to herself. It was her first time by the ocean and, as an attractive young woman of sixteen, she enjoyed the attention she received while swimming at Kitsilano Beach.

Mrs. Davidson helped Ida locate her father's sister, Therese, who was living in Vancouver, not far from Kitsilano Beach. Mrs. Davidson agreed that Ida could spend her day off with her aunt each week during their stay in Vancouver. When Ida was visiting her aunt, she met an Italian man, Angelo, who seemed to be present every time she came by and tried to woo her. Although Angelo was persistent, Ida would not go out with him. She had no intention of breaking the vow she had made earlier.

One day, Angelo did drive her back to the Davidsons' cottage in his nice car. As she was getting out, he gave her an expensive watch — a beautiful thing, covered with sapphires and diamonds. She accepted it without any comment and quickly ran into the cabin. In her mind, the image of her Casanova father giving her mother an expensive ring contrasted with the way he treated her in the following years.

After she returned to Drinnan, Angelo continued to try to contact her by mail, but when Ida told Mary about meeting him, her sister exploded.

"Angelo? Ida, don't you dare have anything further to do with him! He is a scoundrel! He tried to take me out when I visited Auntie. He is older than me, and way too old for you. So stop it!"

Mary never elaborated on the reasons for her violent reaction to the man. Ida did not reply to any of his letters. Soon, Angelo was merely a memory and fodder for family stories.

Ida loved to ride horses and she soon met a Métis family, the Moberlys, whose nearby ranch was home to many fine horses. She became friends with Ed, who was also seventeen, and his younger sister, Sarah, who was Stella's age. Ed was a handsome young man with the rich colouring of his Métis heritage, and Ida had a bit of a crush on him.

One day as they were riding over the rolling terrain, he told about how one of his ancestors, Henry John Moberly, had been the Hudson's Bay factor at Jasper House from 1858 to 1861. He married Suzanne Kwarakwante, an Iroquois, and they had two sons, Ewan and John. Their descendants helped in surveying and opening up the area. When Jasper Park became a federal preserve in 1907, the Canadian government legislated against privately-owned land within the national park boundaries, and so officials had moved the family to a large ranch near Hinton. They ran trapping operations and guided for the outfitting industry.

The Linteris girls loved to get out and ride across the countryside with the warm sun toasting their Italian cheeks, and Ida's face turned to a bronzed brown each summer. One day, with her pant legs tucked into her socks and wearing a headband to hold her hair out of her eyes, Ida invited Stella to join her and they set off for the Moberlys. Stella looked at her for a moment.

"You know, Ida, you look a lot like the Moberlys. You'd fit right into their family."

"Don't be so stupid, Stella. I can't help it if I tan better than anyone else in the family. Anyway, maybe it would be easier in this town to be an Indian than an Italian."

While Ed was great with horses, he lacked any dancing skills, so Ida agreed to teach him. They began going regularly to Mary's house to dance to her phonograph records. It was not long before Ed and his young sister, Sarah, joined the gang at the Saturday night dances that alternated between Hinton and Drinnan. Sarah was only allowed to go to the dances if her father, John, who kept a very watchful and protective eye on her, attended. He carried a flashlight with him and during the 'moonlight' waltzes with all the lights turned off, his flashlight would spotlight his daughter and her partner as they moved around the floor.

In the winter, skating at Thompson Lake also provided many social hours, with the whole gang using shovels to clean off an area, then lighting a bonfire to keep warm. The social interaction between the sexes was always in the context of a group, yet Frank began to wonder about the amount of time Ida was spending with Ed. He told her that he did not want any Indian grandchildren. Ida assured her family that Ed was just a good friend and they had no romantic involvement — but if they did, no one would stop her from going with him or anyone else.

Chapter Thirteen
And the Two Shall Become One

Jim knew that Jock Davidson would welcome him back.

He picked up his duffel bag and skirted the Calder yards until he was in the safe area west of the city. Once again, he found an empty boxcar and drearily settled back for the long lonely ride. He was hungry, dirty and dismal as he jumped from the train when it stopped for water at Pedley just before reaching Drinnan. He began the two-mile walk along the tracks to the bunkhouse, thinking that he just wanted to have a shower, get something to eat and fall into bed. He was disappointed to see a couple of people walking toward him — he didn't want to explain why he was back. As they neared, he saw the beautiful black-haired girl he had seen at numerous dances and her cheeky young sister. He threw his shoulders back, readjusted his bag, rose to his full height and pressed forward into his future. He watched them as they approached, regarding the beauty of the tall dark one.

"Hi," Jim said brightly.

"Hi," Stella quickly replied.

Ida silently gave the dirty tramp a quick once-over.

"Where are you going?" Stella queried.

As the three of them stopped, Jim flipped his duffel bag to the ground.

"Och, I'm warm." He drew the back of his hand across his dripping forehead and continued, "To Drinnan."

"What for?" Ida asked.

"To wait until the mine starts up."

"Where've you come from?" Stella continued.

"I'm just back from Toronto."

Ida had been distracted by his pack and unkempt clothes and had taken him for a hobo, already a common sight during those early Depression days. However, as he spoke and the Rs rolled off his tongue, she took a closer look at the smiling brown eyes glistening above the scruffy beard.

"You're that English miner who likes to dance?"

"Scottish," Jim corrected her. "But you're right about the dancing, lassie. And I remember you. You're one of the Moberlys, aren't you?"

"And what if I am?" Ida spat back.

"Nothing . . . It's just that I've seen you all dancing together and you're good."

"We're Italians, not Indian," Stella interrupted. "I'm Stella Linteris and this is my sister, Ida. What's your name?"

"Jim," he said quickly. "I didn't mean . . . uh . . . that . . . I thought . . . you . . . och . . ."

"The mine won't be up for at least four months," Ida cut him off before the usually-confident Scot could find his words.

"I know, but the manager told me before I left that I could batch in the bunkhouse 'til it opens. I worked there last winter." With a grunt, he flipped his bag onto his shoulders. "Well, I'm off."

He continued his slow walk toward the town. The girls continued on their way for a few steps.

Suddenly, Ida said, "We should be getting back."

They turned to head in the same direction as Jim. With Stella urging Ida to speed up and Jim carrying his heavy duffel, they soon caught up to the Scotsman.

"So, you're gonna live in the pest house," Ida said.

"What's that? Jim replied.

"Oh," she quickly said, a little embarrassed as she realized how her remark had sounded. "It's the nickname some people have for it 'cause the guys don't keep it very clean."

Jim smiled as he looked at Ida. Stella was walking on the rail with her hand on Ida's shoulder, her attention focussed on her balance.

"What we need," he joked, "is a woman to tidy up for us. Would you like to do that?"

"Ida works for Mr. Davidson," Stella answered.

"Jock? He's a good man," Jim responded. "Brought me up here from Clover Bar when he came to manage the mine."

Ida didn't want to talk about her job or the Davidsons, but she was becoming more interested in this handsome man.

"Do you play ball?" she asked.

"Aye, football."

"No. I mean baseball. We need more players on our team."

She began to explain more about the game and that they played against Hinton on Sunday afternoons and really had a great deal of fun.

As they walked, the conversation became more of a dialogue between Ida and Jim. Stella worked hard at stifling the excitement she felt as she watched her older sister talking to this good-looking stranger, and busied herself picking up pebbles and tossing them at the rails ahead.

"What did you do in Toronto?" Ida asked.

Jim welcomed the opportunity to talk of his trip and began regaling the young women with stories, embellishing them in an attempt to impress Ida.

The trio arrived at the Drinnan road, and the sisters realized it was time to part ways.

"We live a little farther on," Ida said. "I hope you'll come out to the ball field on Sunday."

With a bright "Cheerio," Jim turned toward the town and the two girls continued along the tracks.

"We'll bring an extra sandwich for you," Stella hollered after him.

He turned and waved.

Stella turned to Ida and finally blurted out what she had been holding back.

"I think he is really taken by you."

"Don't be so silly, Stella," she offered weakly.

"I hope he joins our ball team."

"We'll see."

Jim did show up for the ball game on Sunday and initially offered little more than enthusiasm, as he had no idea how to hit, throw or catch a baseball. He had seen cricket matches in the old country but that was the closest he had ever come to a game like baseball. However, he soon caught on, and found that his running speed was to his advantage when he chopped the ball into the infield dirt and beat out a throw to first. However, he was there for more than baseball.

On that first day, he asked Ida if he could walk her home. She agreed, but to his chagrin, her sister and her two brothers accompanied them. Jim was frustrated because Stella stayed close to Ida in order to hear what they were talking about, while Pete kept asking Jim about life in Scotland. In spite of this lack of privacy, which continued in following weeks, Ida and Jim got to know each other better week by week.

Ida worked from seven to seven each day, except Sundays. She ate dinner with the Davidsons each evening and when she finished cleaning up the dishes, she headed for home. Her father knew how long it should take her to walk the mile and a half from the Davidsons and gave her strict orders to come right home. However, her route took her past the bunkhouse where Jim and the other single miners lived, and it was not long before the couple fell into a pattern.

She would whistle as she neared, and Jim would come out and walk her home. Since he had eaten his supper, he was able to hang around the front porch every night, talking until her father decided it was bedtime. Frank did not like the idea of an Englisher courting his daughter and was not subtle about letting Jim know this. Every evening when he felt they had been together long enough, he would bellow, "Ida, it's time you were in bed!" Following a lingering goodnight kiss, she would obey her father and Jim would return to his digs, where he entertained the others with stories of his escapades in Scotland.

Many of the miners were emigrants from Europe and since most of them enjoyed playing soccer, they formed a team in Drinnan. Andy Craig, one of the players, had immigrated from Alva and soon he and Jim became close friends. The miners in Hinton also formed a soccer team and soon they were playing friendly matches against each other. In the summer of 1932, the Drinnan team decided to enter a tournament in Jasper. They couldn't afford to buy uniforms, so Nan Thompson, the wife of the mine's pit boss, offered to make shorts for them from bleached flour sacks. Prior to leaving for the game, all the men shaved their heads with straight razors so they would look more like a team. One of the bachelors owned a large touring car and the whole team jammed into it for the trip. Ida was one of the supporters who followed in Jock Davidson's car.

Some American tourists, staying at the Jasper Park Lodge, on hearing that there were coal miners in town, were curious to see what they looked like. During the game, the Drinnan supporters enjoyed the comments from the Americans as they speculated about the miners and their bizarre appearance.

Jim approached Ida at the end of the game to ask what she thought of it. She smiled.

"Who won?" she asked.

As summer turned to fall, Ida and Jim saw more and more of each other, not only on the evening walks from work to home, but also on the ball field and at the weekly dances that alternated between Drinnan and Hinton. These dances were community activities and, although Frank and Angelina did not attend them all, they often went to the hall in Drinnan where Frank was one of the best dancers. He did admire Jim's skill on the dance floor, though he thought him to be second-best. Finally, when he realized that the relationship between Ida and Jim was getting serious, he accepted the fact that she would not be deterred and that he would need to compromise his feelings. He knew, from the previous winter, that Jim was one of the hardest workers in the mine, and was well-liked by the other miners.

He may be a good husband for Ida. But this will not change how I feel about Englishers.

Although the four-mile walk to Hinton was a group experience, Ida's siblings and their friends respected their growing relationship and began giving them more time alone together. Still, as the two strong-minded young people got to know each other better, they often had arguments that ended with Ida storming into the house or Jim fuming on his way home, wondering what to do next.

One fight in particular saw Ida incensed after Jim made an arrogant remark about the 'bohunks' in town.

"I don't think I want to see that man again," Ida said angrily to her brother Pete. "He has no right to call people names. Those Scots think they are so superior."

"Ida, you know you're going to marry Jim, so stop being stupid," Pete replied.

"I won't marry a man who thinks men have all the answers and women are just here to support them. I'll be an old maid first."

"He's not as cocky as he was when he first came here, Ida. He loves you and you love him, so don't cut off your nose to spite your face."

"You sound like Papa, always an answer. But you're right. I do love him and that's what makes it hard when he acts so stupid."

The disagreements did not last long. Soon, the bunkhouse residents heard the whistle again and the grinning Scot bounded out the door to experience the joy of making up. The relationship deepened and the little breakups and reconciliations strengthened the bond between the two. Soon, they were talking marriage.

As Frank slowly accepted the inevitable, tensions eased at home and Jim soon progressed from the front porch into the kitchen, where Angelina was pleased to make tea and serve him her warm fresh bread. On a few rare occasions, if Frank had had a good day at work, he would even offer Jim a drink. The family began to get used to having Jim around, and although he was never invited for dinner, he did spend a number of evenings sipping tea at the kitchen table. Angelina, shy about her lack of English, did not say much, but appreciated how polite Jim was. He quickly grew to admire Angelina and her warm motherly nature.

Shortly after the mines had begun working again in September, Jim decided to ask Frank if he could marry his daughter. Frank looked at him, then at Ida.

"What are you going to do about religion?" he asked.

Ida stared at her father. *What does he care about religion?* She and Jim had talked at length about that because, not only did he have a conservative Baptist upbringing, he had also been inculcated with a strong anti-Roman Catholic bias and constantly defended his position. *I hope Jim doesn't make Papa angry.*

"Let *us* work that out," Jim replied simply.

Frank looked him over again, as if seeing him for the first time, and weighed the pros and cons of having an Englisher in his family.

"You're a hard worker and will provide well for my daughter," he grudgingly offered, saying no more.

"Thank you, sir," Jim said.

They warmly shook hands.

"Pete," Frank called out. "Go down to the root cellar and bring up some of my homebrew. We need to celebrate. Angelina, cook us some food."

Frank knew how to party and his homemade beer did wonders to help Jim really enjoy the strange Italian pasta as they celebrated the coming marriage.

Jim had paid off his bill for groceries at the company store and saved a few dollars to buy lumber. He began building his own house on the lot next to the Thompsons. Never having done anything like this before, he enlisted the help of a delightful black man known as Snowball, who had come from the American south to work in the mines and had construction experience. Frank and Carl also pitched in. Within two weeks, the house began to take shape. As there were no secrets in a mining town, the word was quickly out. Jim and Ida were going to be married.

One morning, as Jim was preparing to go underground carrying a bag of freshly sharpened picks for the coal-cutting machine, he ran into Jock.

"Jimmy, what's this I hear about you getting married?'

"Aye, to Ida Linteris."

"You're going to marry a Papist wop?"

"What the hell did you say, you crazy bugger? You're not going to talk about the woman I love that way!"

With that, he started to throw the sharp steel picks at Jock.

"Stop it! What are you doing, Jimmy?" screamed Jock.

He fled toward the mine office.

"You'll kill me if you hit me with one of those things," he cried.

"If the picks don't do it, I'll break your bloody neck when I catch you," Jim yelled.

He took off in pursuit, all the while throwing picks.

"You're just worried that you're going to lose a good housemaid."

"Stop it, Jimmy! I'm sorry. Just calm down."

Jock reached the safety of the office. Some of the men waiting to go on shift helped mollify Jim enough for him to hear the repeated apology being offered by his boss. The others were concerned that they might lose the hardest-working miner they had ever seen, as there was no doubt in their minds that he would be fired over this. However, this was not to be. When both men had cooled down, Jim accepted the apology. Jock knew by the pay slips that Jim's output was greater than most miners. There was no way he could fire him. He also knew that he was wrong to have spoken about Ida that way.

It was still on Jock's mind when he arrived home for lunch that day and saw Ida.

"I sure made your boyfriend mad at me," he chuckled. "You'd better watch out for him. He has a real temper."

He did not go into detail, so Ida had to wait until that evening to get the full story from Jim. She affirmed his hunch that Davidson was concerned about losing his housekeeper just before his wife was due to have their second child, but the incident also stirred a deep hurt that Ida had been carrying for years.

"It's not fair, Jim. The way you get treated and how awful it is for others, like my father."

"What do you mean, lassie?"

She told him about the incident when her father was threatened with being fired if she did not give up the top student award.

In late October 1932, Mary Davidson went to Edson to have her baby, and Ida had to take over the cooking. When the stakes Jock had requested for himself and the mining engineer turned out to be tough and charred, he was not impressed.

"I have a good mind to fire you, Ida. You are no good as a cook," he told her the next day.

"You can't fire me, Mr. Davidson, because it was your wife who hired me, and that's that!"

"You're pretty saucy."

"Maybe so. But I have learned that it is important to stand up for yourself. Besides, I'll be leaving soon to get married."

"Don't remind me of that."

"My friend Bunty Thompson is willing to take my place and I can train her for the job before I get married if you want me to. It's up to you."

Bunty's father, Jimmy, was the pit boss at the mine, and Jock knew the family. He also admired Ida's spunk.

"All right, but make sure you teach her well."

"She is really smart, Mr. Davidson, and I know she will do a good job."

Now Ida not only had to continue cooking for the family but also had to teach Bunty. In spite of her lack of experience, she did her best to prepare meals that somehow kept the family happy, and Jock managed to hold his tongue during some tough-chewing meals.

The bedrooms in the Davidsons' house were upstairs, and with no indoor plumbing, Jock's father, who lived with them, installed a thunder-jug under his bed. One day when the family was out, Ida was showing Bunty some of the basics. The task at hand was emptying the vessel. But Bunty tripped at the top of the stairs, spilling the smelly fluid down the stairway. After giggling their way through mopping up the mess, they became aware of how the pungent urine smell was permeating every corner of the house. They quickly found a stash of lavender oil and began sprinkling it everywhere. It did disguise the offensive odour, but it was almost unbearable in its sweetness. When the family returned a little later, Ida explained that they were trying to make the house smell nice and that their son, Jackie, had got into the lavender and spread too much of it around.

Ida also taught Bunty how to handle the overbearing Mr. Davidson. They nicknamed him King Tut, and would often refer to him that way when they were caring for young Jackie. This practice stopped the day that Jackie looked out of the window to see his father coming home for lunch and said, "Here comes King Tut."

One morning, as the two young women were returning from the company store carrying a twenty-pound bag of sugar between them, Ida had a thought.

"Let's go and see how Jim is doing on the house."

"Why not?" giggled Bunty, "He'll be surprised to see you."

Jim was working alone when he heard the voices and the laughter of the girls. Peeking around the building, he was shocked to see his young bride-to-be skipping along with Bunty, swinging the big bag between them.

Good God, this is no way for a woman who is about to be married to act. Jim's old-country sense of decorum expressed itself in deep embarrassment — he quickly slid into the crawl space under the floor and hid until they left. He would not relate this to Ida until they had been married a few months.

Once they had set December 2nd as the date for the wedding, Ida arranged to have a neighbour, Mrs. Amos, make the wedding dress. A skilled seamstress, she soon created a beautiful cream-coloured gown. As she sewed, she was free with her advice to the innocent young bride-to-be and gave her a few helpful hints, including tips that verified her reputation in the town for regularly smacking her husband to keep him in line.

With Snowball's experience and Frank and Carl pitching in, the new two-bedroom house, complete with pantry and root cellar, was ready for occupancy by mid-November. Jim moved in. Soon he and Ida found an oak dining room suite for $30 — table, chairs and buffet. They also bought a stove, Winnipeg couch, bed and dresser and were all set up for life together as husband and wife. They had used all of the money Jim had saved, and although they had no debts, they knew they would begin life together living paycheque-to-paycheque.

Jim was adamant about not being married by a Roman Catholic priest. Although Ida had attended infrequent masses over the years and knew her mother might be upset, she agreed with Jim. Since there was no justice of the peace in Drinnan, Jim arranged for a couple who owned a car to drive them to Jasper. This meant that no one else could be present for the ceremony as it was too expensive for the family and friends to travel that far in winter.

The day before the wedding, a huge snowstorm hit the area, blocking the road and their planned excursion. The only choice was to take the daily train to Jasper in the morning and return on the evening train. Jim checked his wallet and figured that he had enough money for the train tickets and a meal in Jasper and he hoped the justice of the peace would not need to be paid. Frank gave Ida a few dollars so that they might buy wine for the wedding celebration. They packed a case big enough to hold Ida's newly-made wedding dress and set off, with Ida reminding Stella that she and her friends had promised to clean out the bunkhouse while they were gone. When they arrived in Jasper on that Friday, December 2nd, 1932, Jim asked the stationmaster how to find the justice of the peace.

"Ye'll nae find yun here, laddie."

Not the reply he was looking for, but he certainly understood the language.

"We have to get married," Ida said quietly to Jim. "My dad will kill me if we spend a whole day together and go home without being married."

"Can you help us out?" Jim asked the man, who turned out to be from Tillicoultry.

"Sure, there are a few kirks in town. I can tell you how to get to them, laddie. Dinna worry, lassie," he told Ida, "it will all work oot."

"We don't want a priest," Jim replied.

By default, they ended up at the United Church manse where the minister welcomed them in. After hearing their story, he called to his wife who took Ida into a bedroom to change into her wedding dress.

The minister's wife and a neighbour witnessed the event, and then, after directing them to a little café where they would have their wedding dinner prior to catching the train home, they all wished the newlyweds well and sent them on their way.

Walking along the snowy road from the manse, Jim smiled and took Ida's hand.

"This will be our church from now on," he announced.

In reality, over the years, the 'our' became everyone else in their family, as Jim said he'd had enough religion as a child to last more than a lifetime.

They went out for their wedding supper to a tiny café, where Jim felt it was appropriate to honour his bride's Italian roots and order the only thing on the menu that might fit the bill: macaroni and cheese. They topped off the meal with Boston cream pie.

After the meal, they took Frank's money and bought three gallons of the best red and had $2.45 left over. Then, loaded down with their wine, they caught the train back.

Unfortunately, Jim did not have enough cash left to grease the conductor's palm with the customary five dollars, so he would not stop the train for them at Drinnan. They had to stay on for another four miles until the regular stop to take on water at Pedley. The wine was too heavy to carry four miles so they stashed it in the snow and tramped toward their new home. When they left the relatively snow-free rail line to skirt the town, Ida did not want folks to see them trudging through the waist-deep snow, so they circled past their own house and around behind the mine office to the home of their neighbours, the Thompsons. Bunty and her family helped them dry out and gave them something warm to drink.

The next day, Saturday, Frank cooked a big chicken dinner for his family and all Jim's friends from the bunkhouse. The only unhappy people at the celebration were Pete and Johnny, who had to pull a sleigh the four miles back to Pedley to pick up the wine from its cache in the snow and drag the load home.

After the dinner, the celebrants went down to the freshly-scrubbed bunkhouse for the wedding dance, with music supplied by a pickup orchestra of local miners playing an accordion, guitar, violin and banjo. The whole mining camp — men, women and children — showed up for the party. Most of the wedding gifts were for use in the home and all had come from the Eaton's catalogue and been delivered by train from Edmonton.

When they arrived home after the party, the newlyweds snuggled down into their new bed and marvelled at how fortunate they were for, as tiny as the house was, it had more space for the two of them than either had ever experienced in a home before. It was very posh in its simplicity.

Jim and Ida's first argument following the wedding set the tone for their marriage. When Ida began to make the bed on the morning after Jim's first shift at the mine, she wondered why the sheets were so dirty and decided to wash them. It was not easy and, as she vigorously scrubbed the sheets on the washboard, she repeatedly rubbed the bar of Sunlight soap into the dark stains. When they were rinsed and white once more, she bundled up against the cold and, with numbing fingers, pinned them to the clothesline, where they soon stopped steaming and began to freeze. Later, she took the board-like sheets indoors and hung them to dry on a line that went from one wall of the kitchen to the other. It took all day for the process, but by the time Jim returned from work, she had remade the bed and the moose stew was simmering on the stove.

However, the next morning, it was dirty sheets all over again. This time she didn't wash them, but left the unmade bed for Jim to see when he got home from work.

Jim liked to eat promptly at five when he returned from his day shift at the mine. As they sat down to eat, Ida poured herself a cup of tea and placed the cozy over the pot to let it steep long enough to satisfy Jim's desire to have it very strong.

She said nothing about the sheets until they began to eat.

"Are you washing at the mine?" she asked.

"What do you mean?" Jim replied.

"The sheets are filthy black every morning. It must be coming from you."

"Well, I shower every day," he replied. "Maybe it's from my back."

"What do you mean your back? I know it's hard to wash it, but the sheets are disgusting."

"I don't wash my back! It would take away all my strength," Jim replied, somewhat annoyed that she didn't understand what he was talking about.

"Don't be stupid! Washing your back doesn't make you weak."

"I'm not stupid! Everyone knows that's what happens."

"Where did you learn such nonsense?"

"It's true! I was taught that as a child. My father never washed his back and neither did any of the men in the pits back home."

"Well, my dad always washes his back and he is not weak. Let me see your back."

"You're not going to wash it!"

"Jim, it can't be good to have all that coal dust in your pores."

"My sisters rubbed me down with a fleece after every shift and that was good enough."

"I don't care about your sisters. I'm not doing any more sheets until you wash your back. In fact, if you want to sleep in that bed with me, you must wash your back."

"You're making me very angry, lassie!" Jim yelled.

"Don't yell at me!" she screamed back. "Just let me see your back."

Breathing heavily, Jim reluctantly unbuttoned his shirt and slowly took it off. Then he undid the buttons on his long johns top and slipped it down over his shoulders.

Ida got up and stood behind him.

"My God, Jim," Ida exclaimed, "it's filthy and it looks like it has blackheads all over it. You've got to let me wash it for you. Your underwear's all black too."

"But what if they're right? I'm one of the strongest men in the mine."

"It's a superstition. You'll be fine."

Ida poured some warm water from the reservoir at the end of the stove into the basin. Wetting a face cloth, she began to gently wash him.

"You've got some sores here. Doesn't it hurt?"

"Sometimes."

Jim was unhappy about the bathing at first, but his mood changed quickly.

"I have to tell you, lassie, that warm water feels good."

When she had finished a second wash with Lifebuoy soap on the cloth, she gently patted Jim's back with a clean towel. He put on his shirt, leaving the underwear top hanging behind him like an apron. Once more, they sat down to the meal that had cooled as she had bathed him.

Jim spoke first.

"I was so mad at you," he admitted, "I . . . I . . . almost hit you."

"I thought that might be what you were thinking." Ida said. "Did you notice my hand on that hot cup of tea? If you had touched me, you would have had that poured over your head."

"I don't ever want to hit you," Jim said quickly.

"I want you to know this," Ida announced, "if you ever, ever hit me, don't bother going to sleep because when you do, I will get you. I will not put up with what my mother did."

"I promise that I will never lay a hand on you in anger," he said with deep conviction.

His respect for women had developed early in his life at home, where he saw his mother as a hero. She came to mind now as he continued to eat.

"Would you write a letter to my mother and tell her about the wedding?" Jim asked.

"You should write her yourself. My father does the letter-writing in our family."

"Well, I'm not good at it and it would be nice for her to get to know you."

"You write a note and I will add to it."

"We should write to Doll, too."

When he began to work in Canada, Jim had mailed his mother $25 a month, at times even dipped into his meagre savings to keep up this commitment. However, with his marriage and new responsibilities, there just wasn't enough money to support a home and have any left over for his mother. Martha had been upset when the money stopped coming and, blaming Ida, would not answer her letters. Ida however, continued to write to Jim's sister, Doll, who kept them up-to-date on family news.

Chapter Fourteen
Motherhood

"Our first Christmas!" Ida exclaimed as she and Jim awoke on that clear winter day in 1932.

"Aye, this is a braw morning."

"I'll get the fire going," said Ida, leaping out of bed.

"Stay awhile," Jim said, reaching out to hold her back.

"We'll have time for that later. There's plenty to do today."

Jim snuggled down until he heard the rattling of the grates and the crackling of the kindling. He reluctantly dressed and sleepily entered the kitchen.

"What are you doing?"

"Making breakfast. It's Christmas."

"We never did anything special at Christmas back home."

"Didn't celebrate Christmas?"

"Oh, we always went to church. But Hogmanay was the big time."

"Hogmanay?"

"New Year's Eve. It's great sport to go out first-footin' and visit friends and family."

"But it will be good with all of us together at Mama's. Christmas is always so special and Papa is usually happy and helps with the cooking."

"You're making such a fuss."

"Well, don't you like our little tree?"

"Aye, it's braw. The wee porcelain birds make it almost like it was living outside."

"We'll keep them and put them on each year from now on. Didn't you give gifts?"

"No. Barely enough money to feed the family let alone buy gifts."

"You never got gifts?"

"My aunt in Glasgow sent things, but we just had a stocking with some sweets in it."

"Well, this will be better than that. There are some presents under Mama's tree."

The whole day was fun. With everyone pitching into make it a good time, abundant laughter and good spirits filled the air. The celebration was topped off splendidly with a scrumptious chicken and pasta dinner. Around eight, when everyone had their fill and it was time for the married couples to head for home, Margaret and Jean were tucked into bed at their grandparents' house. Ida and Jim joined Mary and Carl on the walk back into Drinnan.

As they neared Jim and Ida's house, Jim stopped and turned to the others.

"It's still early. Do you want to come in for a bit?"

"Sure! I've got something to keep us warm," Carl replied, raising high a jug of homemade wine Frank had given him.

Jim quickly brought the fire in the stove back to life, then the four of them sat around the table drinking wine and eating Angelina's Christmas cake. Jim's drink had always been beer but as Carl kept pouring wine into his glass, he emptied it — as was his custom. Carl, amazed at his brother-in-law's capacity, matched him drink for drink. Soon the two of them were singing "O Sole Mio", although Jim had no idea what the words meant, while the wives enjoyed "the silly fools" and their antics. Mary had never seen the usually quiet Carl so animated.

The next morning, Jim, with his head aching and his stomach rebelling, vowed he would never touch that red stuff again. Ida knew that, once he made up his mind about something that would be his way from then on. Moreover, she was right.

The newlyweds quickly settled into the traditional pattern where Jim did the breadwinning and Ida looked after all of the household chores. However, Ida soon got sick with what she perceived to be some kind of flu. It was hard for her to get up in the mornings and she usually threw up when she did. Jim was worried about her and told Frank about it at work.

The next morning, Angelina walked the two miles to her daughter's home. She found Ida in bed and the house quite cold although the fire in the stove was burning well. Looking around, she discovered that there was light coming in under the eaves.

"It is no wonder you are sick!" she announced. "Jim may be a good miner, but he is not a carpenter. Look! He didn't finish the house."

After making her daughter some tea and toast, she returned home and two hours later showed up with a big bundle of rags. Dragging a chair along each interior wall of the house, she stuffed cloth into the gaping spaces. When Jim arrived home from work, he noticed how much warmer the kitchen was, and Ida related the tale of her mother's visit earlier in the day.

Lying in bed next to her soundly sleeping husband the next morning, Ida could only think about how much she hated being sick. She breathed in the carbolic perfume of the Lifebuoy soap that was his daily cleanse and, while it smelled antiseptic, it did nothing to help her nausea. At least with Jim being on afternoon shift, she could stay in bed longer in the morning and delay the inevitable, but the moment always came when she had to give in to it. Slipping out of bed, she dressed quickly, grabbed her coat and dashed to the outhouse to relieve the pressure that kept pushing the sourness upward from her stomach. Once the heaving had subsided, she walked slowly back to the house, drawing deeply of the cold February air that eased the nausea and calmed her.

The last thing she wanted to do was cook, but she felt she should make Jim breakfast as her mother had always done over the years for her father and as Jim's mother and sisters had always done for him. Going to the stove, she picked up the crank handle and gently jiggled the grates to allow the dead ash to drop into the pan and expose the remaining coals from the well-banked fire. After adding a bit of kindling and more coal, she replaced the stove lid and, as the fire heated the stove top and slowly warmed the kitchen, Ida began breakfast preparations. Soon the coffee pot was burbling and the porridge blipping its tiny volcanoes.

Jacket in hand, Jim stopped by the kitchen on his way to the outhouse.

"How are you feeling this morning?" Jim queried.

"I'm fine. Your porridge is ready. Do you want toast? Mama brought a couple of fresh loaves over yesterday along with the rags."

"Aye," Jim said, dashing out the door. "I'll be right back."

Ida smiled. *He always has to go in a hurry every morning.* She moved the porridge and coffee to the back of the stove to make room on the hottest part for the wire rack. The fresh bread was hard to cut into anything but thick slices, but it didn't matter, as Jim liked Angelina's bread with lots of butter and jam.

"That feels better," Jim said on his return.

He washed his hands in the basin before sitting down to enjoy his breakfast.

Ida had been brought up to be a strong, independent and private person and had learned how to conceal her feelings much of the time, but this daily nausea and vomiting could not be hidden from Jim. He noticed her holding it in as he poured the cream off the top of the milk bottle onto his porridge.

"Ida," he said. "I think you should see the doctor."

"I don't need a doctor. I'm fine. It's just a stomach upset that goes away. Once I've had a piece of toast, I usually feel better."

After two months of marriage, Jim was beginning to learn that it was not much use arguing with Ida and he let the subject drop again.

When Jim was on afternoon shift, Bunty often popped in on her way home from work and the two young women shared stories and laughter.

"How's married life going these days?" Bunty asked, punctuating her question with a giggle.

"It's fine, Bunty. But I have not been feeling well lately. I have an upset stomach almost every morning."

Bunty leaned forward, an eyebrow raised.

"Does it last all day?"

"No!" Ida replied.

"Ida, have you had your monthly?"

"I don't like talking about those things."

"Well, I do, and Mrs. Davidson has been telling me what it was like when she was pregnant. She was sick like you and you don't have a monthly and . . ."

"She never talked to me about such things."

Bunty pressed on.

"Ida. Your monthly?"

"Well, I haven't had one since I got married. I thought maybe that's what happens."

"Oh, Ida, you're going to have a baby!" Bunty exclaimed, jumping up and bounding over to her friend. "You're going to have a baby!"

Ida sat for a moment, resisting as Bunty grabbed her hands and pulled her up from the chair.

"Do you really think so?"

"I'm sure. Get your coat on. We've got to tell my mother."

In those days both the spoken and unspoken question always asked of newlyweds was, 'Are you pregnant yet?' Therefore, Nan Thompson was delighted to confirm Bunty's hunch. As the three women conversed, Ida told them that her mother never talked about things like monthlies or pregnancy or morning sickness or having babies, that she hadn't even known Mrs. Davidson was going to have a baby until two months before it came. She timidly asked Nan what it was like to have a baby.

"Oh, my dear," she answered knowingly. "You will just go to sleep in the hospital and they will place your baby in your arms so that it's there when you wake up."

Ida was waiting for Jim when he returned from work shortly after midnight.

"What's the matter? How come you're not in bed?" he queried.

"I'm fine. I couldn't sleep so I decided to wait up for you. I made a pot of tea."

"You're very lively for this time of night. You'll want a long lie in the morning," Jim said.

He removed his coat and boots and sat at the table.

"Fresh biscuits," he noticed. "You must be feeling better."

"I'm fine. I was at Bunty's tonight and Nan thinks the reason I have not been feeling well is that I am going to have a baby," Ida announced, releasing a few hours of pent-up emotion in the process.

"Good show!" Jim exclaimed. "A baby! Come here." He pulled her into his arms. "I couldn't be happier, lass. Oh, I better not squeeze you too hard."

The news spread quickly through the community and the following Saturday night, when a number of the young adults gathered for a sleigh ride into the country, the congratulations rang out repeatedly.

"Now, lass, you sit up here with the driver," Jim said.

"I'm not doing that!" Ida replied emphatically. "The fun of the ride is piling onto the straw and throwing people off."

"Well, you're a married woman now, and you're going to have a baby. You have to take care of the wean. Here's a hand. Be a good lass now," Jim said with some authority, and boosted her up.

Ida did not enjoy the outing and grew angrier with Jim as the evening progressed. It irked her that he was back there pushing people off the sleigh, throwing snow and leading in the singing, while she sat with the driver facing the back ends of two horses. When they arrived home, she reminded Jim clearly that she did not appreciate anyone who tried to control her activities, and in the future she would be the one to determine what a pregnant married woman might or might not do.

By early March 1933 when the mine shut down until fall, Ida and Jim had only been able to put aside a few dollars. Hearing that there might be work in Princeton, Jim decided to hop a freight train and check it out.

Ida, who continued to suffer from extended morning sickness, moved in with her family. Angelina did not say too much about the pregnancy but was able to offer Ida dry soda crackers each morning to help fight the nausea. It felt good to be cared for and spend time with Stella. The two girls picked up where they had left off a few months earlier and within the week were out enjoying the back trails on horseback.

It was one discouraged husband who returned three weeks later, unable to hide his disappointment that the rumours of work had been untrue.

"It won't be easy for the next while," he said.

As if to demonstrate the point, he rubbed his hands together vigorously. They hadn't been back long and were still waiting for the newly-lit stove to begin to warm their chilled home.

"I'll keep looking," Jim assured. "There's got to be something I can do to tide us over until the mine starts again. How was your time with your family?"

"It was good. Stella and I had fun," Ida replied, hoping that he might get caught up in her joy. "I've got some pictures that Papa took with his new camera."

She reached into her purse and retrieved the photos.

"I wish you had been with us," she said as she handed him the pictures.

Jim shuffled through them, smiling at first, but then shock crossed his face.

"What were you thinking, Ida? Look at you, on a horse!" Holding up another picture, he added, "You and Stella dressed up in men's suits. Good God, woman, have you lost your senses?"

"What're you fussing about? You know I love to ride. We thought it would be fun to put on some of Pete's clothes. You sound like an old fogey."

Once again, an argument ensued as to what was and what wasn't appropriate behaviour for a married pregnant woman. Neither budged.

In May 1933, Frank moved his family to Lethbridge, where the job was waiting for him at the Galt mine. Ida and Jim would certainly miss the family dinners, Angelina's homemade bread and the fun of having Stella around, but once again, Frank had decided for his family.

Shortly after his return to Lethbridge, Frank called on his friend Simeon 'Simon' Fabbi. A hard-working Italian immigrant, Simon had started a small business with two cows, a number of chickens and a vegetable garden. He sold milk, eggs, chickens and surplus vegetables to neighbours in order to support his three young sons, Eugene, Stan and Romeo. His wife, not liking Canada, had returned to Italy alone when the boys were very young.

A frugal, industrious man, Simon gradually increased his herd and when the Union Dairy moved to new premises, he bought that company's old plant and founded the Purity Dairy.

When the Linteris family had lived in Lethbridge before their move to the Coal Branch, Ida had been in the same grade as Stan, and Stella in the same grade as Romeo, who was also part of her first communion class. With that history, it was not long after Frank made contact again that Simon hired Pete and Johnny to help him with his cows and chickens and Stella began to work for him in the dairy. Soon, he convinced Frank to quit working in the mine and take up employment with him at the dairy.

During this time, Sarah Philgate, a widow in her fifties, moved in next door to Jim and Ida, with her recently widowed son-in-law, Jack Forsyth, and his two boys. Because they had no well of their own, Ida and Jim carried water from Jack's well and, in the process got to know and appreciate Sarah, especially after Ida's family moved.

Jim's spirits picked up in the late spring when he heard about a job on a dude ranch being constructed across the Athabasca River from Hinton. He arranged for an interview.

"Do you have any experience as a cook?" the foreman asked.

"Aye, I worked for the railroad for a while in the kitchen car."

"We have a good cook for the construction crew but he needs an experienced helper."

"I'm your man," said Jim, exuding the confidence that he could tackle anything.

After accepting the job, he became concerned that Ida would be alone for a few months as the only access to the ranch was by way of the bridge at Entrance over fifteen miles west, and with no car, he would not be able to travel back and forth. He asked Sarah to keep an eye on Ida and she assured him that it would be her pleasure to do so.

"Don't worry about me," Ida said. "I'll be fine. Take care of yourself."

"I'll miss you lassie," Jim assured her.

He gave her a warm squeeze. Then he hoisted his duffel bag, filled with freshly-washed clothes, onto his shoulder and headed out to meet the foreman for a ride to the ranch.

Sarah saw in Ida the daughter she had lost. With Jim away, she spent as much time as she could checking on her, bringing samples of her cooking and baking. She also saw to it that Jack kept Ida's water barrel filled.

Jim returned in late August in anticipation of the mine reopening.

"I missed you, hen," he said tenderly. "It wasn't easy being so close and yet so far away. I walked down to the river a few times and looked over this way, thinking about you and wondering if I could swim across."

"I'm glad you didn't try, you silly ass. That water is too cold and fast."

"Och, you're getting quite big. I can't get as close as I used to," he teased as he drew her near. "Did you get over the sickness?"

"No," Ida said.

She stepped back.

"Even the crackers don't help that much anymore. One of the women suggested I eat coal to settle my stomach."

"Did it work?"

"Don't be silly! I would never try something like that. Anyway, it will soon be over. But I'm not sure I want to do this again."

"Did you see the doctor?"

"No, I'm fine. There's no need to bother him until the baby comes."

Jim knew it was no use arguing with her but was concerned and wished that she would change her mind. He had learned to bake bread while he was at the ranch and soon had taught Ida the fine art.

"I think yours is even better than you mother's," he said one day after he took a huge bite out of the butter-dripping crust he had cut from a hot loaf.

"Thanks, but I don't think I will ever match my mother's baking."

Although the mine was not working, the company still paid for the weekly visits of an Edson doctor, who met patients at the mine office. Jim dropped by the office two days after his return, asking if the doctor could visit his pregnant wife.

"The doctor has a full schedule today but his assistant, Dr. Hunt, will be able to do it," the nurse in the office told Jim.

"She knows of Dr. Scott, but the other one will do fine," Jim said.

He then gave the nurse directions to his house.

When Ida answered the door, she was greeted with the sight of this unfamiliar man.

"Hi, I'm Doctor Hunt and I understand you are going to have a baby. May I come in?"

"Where is Doctor Scott?"

"He is busy with other people. I'm helping him out on this trip."

Choking back exasperation, she forced a smile. *This is Jim's doing. When he gets home . . . It would be rude to send a doctor away.* Without a further word, she opened the door and let him in.

"When was the last time you saw a doctor?"

"I haven't seen one since I got pregnant," Ida replied shyly.

"Well, it's important for you to have a check-up. I want you to go into the bedroom, take off all your clothes and cover yourself with a sheet so I can examine you. Let me know when you are ready."

Ida silently obeyed. Before she had time to question what was happening, she was examined and pronounced healthy.

"Everything looks fine, Mrs. Elliot. We want you at the hospital in Edson in plenty of time. Let's see . . ." He paused to look at his calendar, "There are three trains a week. You'd better come on Monday, September 18. Okay?"

Edson was about fifty miles away along the Canadian National Railroad line. Ida, of course, didn't see this as any great challenge.

"Yes, that will be fine for me."

"Good. I'll let myself out."

Ida lay there for a few minutes thinking that this had been quite amazing. This stranger had come in and examined her and she didn't even feel embarrassed. When Jim arrived home, Ida recounted her experience with the doctor proudly.

"But," she added, "don't you ever do anything like that to me again! I don't like people doing things behind my back."

The force of the statement gave Jim no room or desire to argue.

"I'm sorry," he said.

Monday, September 18, was a big day for the people of Drinnan. The Davidsons were leaving. Jock had accepted a position as manager at the mine in Coalhurst and had decided to go by train with his family, leaving his car for Andy Wilson, the new manager, until Andy could purchase one. Jock had been a popular manager and most of the townspeople planned to gather at the station to say goodbye.

Ida was not happy when she discovered the Davidsons' travel plans.

"I'm not going on that train with all of those people around," Ida declared firmly to Jim. "I don't want a crowd around when I leave. I'll go on Wednesday."

And that was that. She stayed home and Jim went off that afternoon on his fourth shift since the mine had reopened.

No one had explained to Ida just what labour would be like so she did not know what was happening the next morning when she awoke with severe back pains.

Jim noticed her struggling to get dressed.

"What's the matter, lass?" Jim asked.

"Nothing much. I've been having sharp twinges in my back every now and then. It'll be okay once I am up and about. It's already feeling better."

The pattern continued throughout the morning. Although Ida didn't complain as she stoically went about her chores, each spasm jolted her to a brief stop, her face betraying the reality of the pain.

"Is there anything I can do for you?" Jim asked.

"No, it passes. Stop hovering around me."

Jim knew he couldn't stay home from work, as he needed to get in as many shifts as possible. *I'll talk with Sarah*, he thought.

"I'll fill up the water barrel before I go to work," he said.

He picked up the pails and headed for the neighbours'.

Sarah answered his knock and opened the screen door.

"Hello, Jim, how are things today?"

"Ida's not good. Would you mind looking in on her when I'm at work?"

"What's the matter?"

"She's been having back pains on and off this morning."

"Back pains? What are they like?"

"She says they come for a while and then it gets better and then they come again."

"Oh, Jim! The baby's on its way! Get down to the mine office and have them call the doctor in Edson and tell him that we are coming. I'll go over and help her get ready."

Jim dashed home with two half-filled buckets, wet pant legs, and bursting lungs.

"Sarah says you're getting ready to have the baby," he forced out. "She'll be over in a few minutes to help you. I'm off to the mine to get the car."

Not waiting for a response, he gave her a quick kiss and ran off toward the mine.

"Johnny, the baby's coming. Phone the hospital, get me an advance on my pay, find Andy Wilson and get the car gassed up."

Johnny Farmer, a shy eighteen-year-old who worked in the mine office, had been given the keys to the Davidsons' 1929 Chevrolet in case there were any emergencies in town. He listened intently to the fast-fired orders.

"I'll do it right away, Jim. How's Ida?"

"She's doing fine."

With that, Jim ran back toward his home, trusting everything would be done as he wished. Andy Wilson found men to fill in for Johnny and do

Jim's shift that afternoon. In the process, word quickly spread through the community that Ida was on her way to have the baby.

It was almost two that afternoon when they were finally ready to leave for Edson. It looked as if there were as many people out to see them off as there had been the day before when the Davidsons had left. Jim sat in the front with a blushing Johnny beside him. Ida, embarrassed by all the attention, sat in the back seat beside Sarah, who had brought an armload of towels that were piled high between the two women. Johnny carefully drove the car out from the town and on to the road that was simply the abandoned Grand Trunk Railway Line with the tracks removed.

They had hardly settled in for the trip and were not quite out of town when Johnny got a strange look on his face.

"There's something wrong. I can't steer the car properly. I think we have a flat."

"Do you know what to do?" Jim asked.

"Sure, Mr. Davidson taught me all about flats."

Jim opened the door for the women.

"Are you okay, hen?"

"I'm fine. Don't worry about me."

"Come, dear," Sarah said to Ida. "Let's get out of the way and sit on that log in the shade. How are you doing?"

"There doesn't seem to be as much pain now," Ida replied.

Sarah spread out a couple of towels to sit on. Ida was very calm, still not fully understanding that she was in labour. She and Sarah chatted quietly while the men changed the tire.

Once underway again, Johnny kept looking in his rear-view mirror trying to reassure himself that his passenger was all right. There was little conversation as the car rattled and bounced over the rough gravel roadbed. About five miles from Drinnan, another tire went flat.

"Bloody hell!" exclaimed the anxious Jim. "Sorry," he quickly said over his shoulder to the women in the back. "What do we do now, Johnny?"

"I have a patching kit and a pump. Mr. Davidson showed me how fix tubes," assured Johnny, his voice betraying his lack of confidence.

As the two women walked along the road to stretch their legs, Ida talked about how she was looking forward to being a mother. In response to constant questioning, Ida assured Sarah repeatedly that she was doing fine. Sarah admired her spunk. She squeezed her arm and drew Ida's attention as she finally felt ready to ask the question she had been formulating for some time.

"If the baby is a girl, will you name her after me? As you know, my daughter is dead and I only have two grandsons."

Ida smiled weakly. *'Sarah'? I really don't like that name. I hope I have a boy.*

"I'll talk it over with Jim when we get to the hospital," she replied.

"It would mean so much to me," Sarah went on. "Oh, it looks like the men are about finished."

After they had settled in for the third time, Sarah asked, "Are you comfortable, Ida?"

"I'm all right. This road is sure rough, though. I'm glad you brought those towels for a cushion."

Only then did Sarah realize that Ida did not know why she had brought the towels and hoped she would not need to discover the truth. Johnny, overhearing this, got more anxious as they endured the rough road.

"What if she has the baby in the car?" he whispered to Jim.

"Don't worry, laddie, that baby's going to wait until we get to Edson."

"How much longer do you think we'll be?" Ida asked from the back seat.

"We're almost at Vickerdale. So we're about five miles from Edson," Johnny said. "Looks like there's someone on the road ahead."

As they drew closer, they could see a woman frantically waving for them to stop.

"Can you give me a ride to Edson?" she asked excitedly. "My son has been poisoned. They took him to the hospital but didn't have room for me in the car."

"Jump in. We're heading for the hospital ourselves," Jim said.

He got out and opened the back door for her. Sarah patted the woman's arm.

"Don't you worry, we'll be there soon."

The five-hour driving ordeal ended when the crowded car pulled into the hospital parking lot at seven that evening.

With a quick thank-you, the hitchhiker ran off to find her son. The others, relieved that Ida's baby was still waiting to make an appearance, went to the admitting desk.

As soon as Ida was settled in bed, the doctor entered.

"We were expecting you yesterday, Mrs. Elliot, but I'm happy that you made it before the baby came."

He examined her and determined that the baby was still not quite ready to make an appearance.

"It looks like it will be some time before the baby comes," the doctor told Jim in the waiting area. "You can go and see your wife now. There are a couple of good restaurants in town if you want to go for supper. I'll be back a little later."

"Are you okay, hon?" Jim asked, as he stood beside her holding her hand.

"I'm very comfortable and the pain is not too bad when it comes. Why don't you take Sarah and Johnny out for something to eat? I'll see you later."

After supper, they returned to the hospital. Sarah asked Jim if she might have a moment with Ida.

"Your husband is so thoughtful," she said to Ida. "He tried so hard to look after me, paid for my supper. He even asked me if I needed to go to the bathroom."

Ida smiled for a moment and then sobered up on hearing Sarah's departing comment.

"I hope it's a girl."

Jim came into the room. Ida motioned him to come close.

"Jim," she whispered," she wants us to name our baby after her. I don't like the name."

"Well, you'd better have a boy then, eh?"

Ida was gripped by another spasm. Jim looked at her quizzically.

"Is it worse than a toothache?" he asked innocently.

He did not know much about the birthing process as back home everyone was ushered out of the house when a baby was due.

"I don't know," she replied, just as innocently. "I've never had a toothache."

Around nine, the doctor came in and suggested that the baby would be a while yet and that Jim might want to go back to Drinnan.

"What do you think, Ida?" Jim asked. "We could use the money."

"I'll be alright. Don't worry about me."

"You're a brave lass, Ida. I may not get back before the birth."

"That's all right. You go and work hard and I'll do okay."

He reached into his pocket and pulled out some money.

"Here, Ida, this should be enough for the train ticket and something to eat when you come home."

Within an hour, the pains got worse with very little let-up between them. Ida was not sure just how much of this she could take but was determined not to let the nursing sister know just how severe it was. The nun kept encouraging her to hang on and to push.

Motherhood 183

"You're a strong young woman. You won't have too much trouble with this. Just push when I tell you to." The Edson hospital was a Catholic institution and most of the nurses were nuns. For Ida, it brought back memories of her time with the strict Mother Superior at the school in Lethbridge. She would show them how strong she was.

Ida gritted her teeth, clenched her fists and squeezed back tears as she followed the nun's coaching until, with one final push, the pain released and the baby was born — barely two hours after Jim had left.

Her firstborn announced his own arrival with a loud cry.

"You have a fine-looking son, Mrs. Elliot," the doctor announced.

My God, a son. I am a mother!

Chapter Fifteen
And Baby Makes Three

In the early morning hours of September 20, Jim returned home from Edson to find a note from Jimmy Thompson pinned to the door, asking him to work a double shift.

He slept the few short hours until the alarm clock heralded the new workday. Then, although not fully rested, he quickly rose, packed a double lunch while his porridge cooked, gobbled that down and headed off to work, knowing he would not return until after midnight. Nevertheless, he was a happy man thinking about the coming baby and the extra pay.

Partway through the morning, Andy Wilson entered the area where Jim was working.

"The hospital telephoned," he said. "Ida's doing fine and so is your new son."

A son! Jim was ecstatic. *He'll be James, like his father, and his father before him and his father before that.*

"Good show, Andy. Now I've a reason to work all the harder."

In spite of the lack of sleep and the double shift, Jim still had some trouble shutting off his mind as he wearily crawled into bed that night, reflecting on what it might mean to be a father to a son. *One thing I know, he'll never go into the mines.*

Ida was confined to her bed for ten long days — compelled to use bedpans and nurse the baby whenever the sisters decided it was time to feed him.

Then one morning, a sister brought the wrong infant for feeding.

"That's not my baby," Ida said firmly.

"Don't be silly. This is yours. Now take him."

"I will not. I'm *not* going to feed that baby."

"Mrs. Elliot, this is your son."

"I know my own baby!"

The nursing sister left in a huff, returned a few moments later, silently handed the correct infant to the waiting mother, then pulled the curtain to give her privacy to nurse the baby.

On the tenth day, the hospital arranged a ride for Ida and the baby to the railway station. Once there, she was alone with her son for the first time. After purchasing her ticket with money that Jim had left, she sat on a bench facing the platform through large windows, hoping she would not see anyone she knew. It had been a long ten days in the hospital and all she wanted now was a quiet trip home. The *clang, clang, clang* of the bell and the hissing of escaping steam announced the approach of the train from Edmonton and filled her with excitement. Carefully, she cradled the baby in one arm, rose from the bench, picked up her bag and headed out the door.

The conductor placed the stepstool on the platform and began giving directions to the few people boarding there. Ida, the last in line, handed him her ticket. He looked at it and at the bundle in her arms.

"New baby?" he asked.

"Yes," she replied shyly. "He's our first and my husband hasn't seen him yet."

"Come with me."

He took her bag, climbed up into the car and led her down the aisle to a seat near the far end.

"Here you are. There aren't as many people going to the coast right now so we have plenty of room."

He put her bag on the seat next to the aisle as Ida moved toward the window.

"Just let me know if you need anything. We'll have you home in less than two hours."

Ida thanked him before turning to settle her sleeping baby on the seat facing her. He looked so beautiful and peaceful. She sat back, looked around at the wood-panelled walls, the leather seats, the fancy light fixtures, the beautiful carpet runner in the aisle. The car was so different from those used on the coal branch line where she had travelled when she was younger.

Soon, two couples on their way home to Drinnan from Edmonton recognized Ida and came down the aisle to see the baby. Ida was grateful when they returned to their seats at the other end of the car prior to departure.

The train whistle blew and the engine began to puff, slowly at first, then more rapidly as it glided out of the station and on its way. As it picked up speed, the steady clackedy-clack of the steel wheels on the rail joints, the staccato of the engine sounding like an old man whistling without making music and the rhythmical swaying of the car combined to mesmerize Ida and the baby. Although he had stirred a little when she first put him down, he was now sleeping soundly and she hoped he would stay that way — she would *not* nurse in public.

When at last the train slowed, her heart raced. *We're almost there.* But looking expectantly for the town, she saw only trees and she sighed as she remembered how trains always filled their boilers at Pedley before the climb into the Rockies. Still four more long miles to go.

"Drinnan!"

The voice of the conductor brought her back once more. He stopped beside her with a smile.

"I'll help you with your bag," he said. "Just wait until the other people get off."

She smiled, thanked him and quickly returned her gaze to the window. As the train slowed, the gray tipple of the mine came into view and then the station platform. *Must be a lot of people going to Jasper today. I see Jim. Oh, there's Bunty and Jean and Mrs. Thompson and Mrs. McLean and Sarah and . . . oh God, there's a crowd there.* She waved timidly.

Jim, on the platform at the front of the crowd, looked intently into each window of the slowing cars. Finally, he spied her through the reflection of the platform in the train window.

"Here she is," he announced.

He waved and smiled up at Ida, walking alongside the car until it came to a halt. Then, he stood by the stairwell.

The conductor, bent down to place the stool on the platform, looked up at the grinning young man.

"Are you the father of that beautiful baby?" he asked.

"Aye, I am."

"Just wait until these people are off and then you can go and get her."

As soon as the other passengers had disembarked and the conductor gave him the go-ahead, Jim was in like a flash. He rushed to embrace Ida, who had just picked up the baby. He stepped back and looked closely at his son.

"He's braw, lassie."

"It's so good to see you, Jim," she said as she nuzzled close. "Why did you bring all those people?"

"I didn't. They came. You know what it's like. When the hospital phoned the mine, Jimmy Thompson was sent to tell me and then he told Nan and the girls and it went from there."

"Well, I just want to go home. The baby will need feeding. Please don't dawdle."

The welcoming group surrounded them, a hubbub of 'oohs' and 'ahhs' and 'can-I-hold-hims' waking the baby. He responded by adding his voice in turn.

"We're awa'," Jim announced to the crowd. "We've got to let wee Jimmy see his new home."

He put his arm around Ida and ushered her along the platform. The chattering group, all neighbours, followed closely behind as they left the station but dispersed as they reached their own homes. It wasn't long before the new parents were home, the smell of fresh bread wafting out of the house.

"Oh, you've made bread," Ida exclaimed.

"Aye, I did that this morning. I couldn't sleep."

Jim opened the draft on the stove and added more coal.

"The kettle will soon be boiling for tea," he offered.

"I'll be a few minutes," Ida said. "I'm going to feed Jimmy. I'll make lunch when I'm finished."

She took the baby into the bedroom.

"I'll scramble some eggs," Jim called after her. "The Thompsons have invited you over for supper tonight while I'm at work. Nan thought you might be too tired to cook."

"That will be nice," Ida replied. "I can't talk now."

She looked down at her nursing child. *Your father is such a gem. Eggs will be good today.*

Eggs were not Ida's favourite, but Jim liked them on a Sunday, usually the only time and the only thing he cooked. Therefore, he felt she could eat them on this Wednesday, which was a special day. Sure enough, Ida seemed happy for the food as they sat down to eat.

"There's some mail there for you," Jim told her. "Your father says he is doing fine at the dairy and that Stella, Pete and Johnny are all working there too. He's a strange one. Men aren't usually the letter writers."

"Mama can't do English, but I don't see why men can't write. Sometimes I think it is strange that I'm the one that keeps in touch with your family."

"That's the way it is, lassie. You do a good job. There's a letter from Doll, also. Mary is still seeing that race car driver, Campbell. He'd better treat her well."

"Well, we have a baby to think of now. Let's worry about treating him well."

Jim, typical of the men of his time, had nothing to do with tending the little one except to announce periodically 'I think wee Jimmy's hungry. Other times, he would wrinkle up his nose and look as though he might vomit as he gasped 'The bairn needs changing.'

Ida quickly realized how exhausting it was being a full-time mother while continuing with her other homemaking responsibilities. Her routine was interrupted day and night by the baby's need for nursing, changing or cuddling. She quickly discovered that when little Jimmy continued to fuss, she could pacify him by laying him on the seat of the rocking chair and rocking it with one foot. She would do this while performing most tasks. This worked especially well when she was doing the relentless laundry, following her mother's rules. Since her diaper supply only lasted two days, and she wouldn't think of washing other clothes with the baby's things, which had to be boiled, it seemed as though she always had her hands in water.

Fortunately, carrying water was men's work and Jim filled the reservoir and the copper boiler on the stove, leaving both pails full before he left for work each day.

"There you go, lass," he would say, "I hope that will do until I get home."

"I can get more, if I need to."

"I don't want you lugging those heavy pails. Anyway, you shouldn't leave the wean alone."

"I can handle things just fine," Ida said. "Now, you get off to work. Have you got your lunch bucket?"

He did, and as Jimmy walked off to work, Ida readied herself for another day of chores.

As soon as the water was hot enough, Ida got at the wash. She emptied the stove reservoir into the washtub and began rubbing the naphtha soap on the most soiled spots, vigorously working the diapers up and down over the surface of the washboard. After she had wrung the water out of each item, she rinsed them in the second tub and then dropped them into the simmering water in the copper boiler on the stove. As she stirred the simmering diapers with a strong stick, she thought of her mother doing the same thing at Mile 40, not only for the family wash but also for all the clothes she did for the men in the bunkhouse. Ida fished the hot diapers out of the boiler and set them in the large basin she used to carry the clothes outside to hang on the line.

On regular wash days, she started with the whites and boiled them, then the dark clothes were washed and rinsed in the same water but were not boiled. The final wash water was used to scrub the linoleum floor which seemed to attract the outside grit very quickly.

Winter added yet another burden to Ida's wearying load when she had to hang the clothes outside in the freezing weather. This first part was not too difficult as the laundry was warm, but later in the day when everything was stiffly frozen, it was painful on the fingers to remove the frigid clothespins, pry the folded edges of the cloth from the line, then carry the stiff and brittle laundry into the house without breaking the fabric. Once she had manoeuvred them through the door, she had to hang them on lines strung across the kitchen until they were dry, or at least ready for ironing. With most clothing being made out of cotton, the 'sad' (heavy) irons were always on the stovetop, ready for use on washday and the day following. Ida spread a couple of old sheets over the oilcloth covering the table and ironed on that. If the clothes were too dry, she sprinkled them with a little water to dampen them and thus create steam when the hot iron was pressed onto the damp fabric, making it easier to remove the wrinkles.

One day near the end of October, when Jim returned from work, Ida met him at the door with an envelope in her hand.

"We finally got a letter from your mother," she said crisply. "She and your father are pleased with the news of Jimmy's birth."

She handed him the thin envelope. He read the contents in silence and then wrapped his arms around Ida.

"It was a long time coming," he said. "She'll want to stay in touch now that the bairn is here."

"I'll write to her as long as she is civil with me," Ida declared.

Doll had described Martha's concern that Jim had married an Italian and a Catholic.

"She can be a stubborn woman sometimes, but she'll grow to like you. Mind you, she'll have a hard time saying that," Jim replied.

Drinnan did not have a store so the residents ordered their groceries by mail from Woodward's in Edmonton. A local drayman met the train each week and did door-to-door delivery, collecting payment and picking up lists for the next shipment. Local farmers provided milk and butter, a poultry producer from Edson came by train periodically selling eggs, but the meat, mostly deer, elk and moose, was supplied by local hunters. Whenever anyone made a kill, he would share his bounty with others. It was the same with fish.

Jim still loved fishing and took full advantage of the Athabasca River and some of the streams leading into it. He usually was very successful and shared his catch with neighbours. However, while fresh fish were available all year, game was not, so Ida learned to preserve meat in jars. Stew was the most obvious result of this processing and, after a year, it became quite boring for her, although Jim never tired of it.

Jean Thompson spent as much time as she could at the Elliots'. Ida was very patient with her as the nine-year-old was losing her hearing, the result of a serious measles attack two years previously. This affected her speech and she had difficulty making herself understood at times, which embarrassed her older sister Bunty who, at seventeen, was becoming interested in men. Bunty had quite a crush on Mr. Willis, one of the two single young teachers at the school, and convinced Nan to invite him to dinner one Sunday. As they sat enjoying their meal, Jean excitedly told the others about the chipmunk she had seen earlier in the day.

"I shlaw a shit mik!"

"Be quiet, Jean!" Bunty commanded.

"I did! I shlaw a shit mik!"

Bunty was mortified and later told Ida that her stupid sister had probably spoiled any chance she had of ever getting married. Ida was not too sympathetic, as Bunty's sense of superiority sometimes got in the way of their friendship.

"She's only nine, Bunty. How would you like to be going deaf?"

Jean spent as much time as she could with the new baby and became a real help by entertaining Jimmy after he began to teethe. He had developed an itchy rash on the side of his face and around his ears and fussed quite a bit. The visiting doctor thought it was eczema and suggested that the baby might be allergic to the mashed foods Ida was preparing. He suggested commercial baby food, so the next order to Woodward's included jars of Gerber baby food.

"I hope this helps," said Jim when the order arrived. "It's twenty-five cents a bottle."

"It doesn't look much different than what I make," Ida replied. "But let's give it a try."

When some of the neighbours heard about buying the baby food, they told Ida that Jimmy probably just had a teething rash that would soon go away. However, the change in food did seem to help a little and they kept it up, in spite of the expense. When the rash disappeared once the new teeth had broken through the gums, it appeared that the neighbours were right. It indeed was eczema, but a teething eczema, and the baby would have outgrown it without the help of Gerber.

Making ends meet in the Depression years was difficult, but Jim and Ida discovered many ways to enjoy life in a coal-mining town. They both loved walking, fishing and visiting neighbours. Their chief social entertainment was in the Saturday night dances that alternated between Hinton and Drinnan. All through that first winter with the new baby, they would bundle him up and carry him to the dances. The four-mile trek to Hinton was always fun. The whole gang from Drinnan went *en masse* along the railway tracks and there was always someone who wanted to carry wee Jimmy. With an orchestra made up of locals and the few pool tables that had been added to the community hall pushed to one side, the floor was filled with younger folk dancing while the older women, those fort-five and older, tended the babies sleeping on the green felt of the pool tables.

Jim was doing well at the mine now, operating a machine and paid on a contract basis. The new manager, Andy Wilson, another Scot, with two university degrees in mining but little practical experience, quickly recognized that Jim was not only an extremely hard worker but a man with a great deal of experience and knowledge of mining. As the mine manager was always at the top of the status ladder for most of the people in mining towns and he and his family usually kept their distance from the others, it was a powerful affirmation of Jim's wisdom when Andy began to seek him out for technical advice on managing the mine.

Andy, a short, swarthy man, had married Olga, a tall, blonde-haired woman of Russian descent. They had two daughters, two-year-old Jean and Doreen, who had been born just a month before Jimmy. With Andy's need for discussion about mining procedures and Olga's desire for social contact, he asked Jim if he and Ida knew how to play bridge. They didn't, but agreed that they would like to learn. Soon, most evenings were spent around Ida and Jim's kitchen table, with Jimmy asleep in his basket in the bedroom and the Wilson children in their beds at home, a half-block away. Because the power was supplied by the mine, there was no electricity in the town when the mine shut down in the summer, so Olga and Andy always left a coal oil lamp burning in the living room window, which could be seen from Ida and Jim's. Andy would run down to the house periodically to check the stove and the children.

Ida loved playing cards but became bored at times with the constant mining talk. She noted as much one night as they settled into bed following an evening of cards.

"I think the only reason Andy comes over is to discuss what he should do next," Ida whispered to Jim.

"We're the only ones they socialize with, Ida. They never go to the dances or anything like that."

"I know. But you know more than Andy does. I think you should be getting more money."

"Well, he gives me as much work as he can. Aren't you enjoying the cards?"

"Yes, we're getting really good. But Olga acts so superior and she's hard to be around sometimes."

"She got a good catch in Andy," Jim said with a yawn. "Bohunks have a harder time than Italians in mining towns, so she thinks she has to try and prove that she's better than the rest."

"Well, if she's not careful, I will tell her off one day. Always talking about how special her fair-haired girls are."

"Ida, let's get some sleep. The bairn will want food early."

He snuggled into a spoon with his wife.

The friendship with the Wilsons continued without much stress until one summer day when Ida was sweeping the front step. Olga came by with her two blonde daughters. After initial greetings, Olga looked at Jimmy bouncing happily in his spring chair in the doorway and commented on his eczema-scarred face.

"Ida, do you know your baby looks like an African?"

Ida did not think out her response. It came from some deep place of many rejections and much discrimination.

"At least," she blurted out," he doesn't have a square head!"

Olga did not reply but turned abruptly and strutted away. It was a long time before the two couples played bridge together again.

Later in the day when Ida was visiting Bunty and her mom, Nan was aghast that Ida had spoken to the manager's wife that way.

"Ida, you can't do things like that. Jim can get fired. You should apologize to her."

"She should apologize to me! Andy won't fire Jim — he needs him to run the mine. Besides, he's Scottish! I'm not going to let people put my family down."

When Ida received similar advice from another neighbour, Mrs. McLean, she gave her the same reply.

As winter turned to spring, the orders for coal decreased as did working days at the mine. There was enough money to live on but they just couldn't get ahead.

Meanwhile, Jimmy Thompson had taken a job at the Black Diamond mine near Edmonton so that Jean could go to the school for the deaf there and learn how to lip-read. In the spring, Nan wrote that Jean was not doing well. She seemed to have a lot of allergies and her eczema was much worse, so Ida and Jim suggested the girl might spend the summer with them.

When Nan brought her by train, she warned Ida that Jean could not eat certain foods and that she would have to be watched closely. But Jean had a great time that summer, ate whatever was put on her plate and her skin cleared up. She loved the baby and was happy to amuse him while Ida went about her chores.

One idle summer afternoon in 1934, Ida noticed an odd look on Jim's face.

"What's the matter, Jim?" she asked.

"Och, it's nothing," Jim shrugged.

He dipped a bun into the puddle of molasses on his plate and bit off the soggy part.

"Something's bothering you. You've been awfully quiet lately."

"I'm fine, but it's not the way I wanted things to be when I left home. I want wee Jimmy to have a better chance than I did."

Ida took a sip of tea and cradled the cup in her hands as she gazed over it at Jim.

"We've not done so badly. We have a house and a baby," she reassured him.

"I'm not complaining about you or the bairn."

He took another bite of bun.

"We could move somewhere else if you find work," Ida offered, always optimistic about moving and seeing it as a new adventure.

"But our house is here and mining is all I know."

"Mary's always writing about how steady the work is in Mountain Park. Maybe you could get a job there."

He paused and looked around the room, the labour of his hands.

"It's the moving, I guess. Leaving our first home."

"We'll build another someday."

"I guess we could sell this place and have some money to get started there," Jim sighed. "Better write her and ask Carl to see what the possibilities are."

The mine in Mountain Park was supplying coal to the railway and working more than most in the area, so Mary's quick reply was encouraging. She wrote that Carl had talked to the pit boss and he assured him that there would be a job there for Jim.

"Mountain Park has so much to offer," Ida said after they had read the letter. "And it will be a good place for raising children. There's a school, a hospital, a theatre and a couple of stores. The McLeod River is full of fish."

When the time came to move, they packed up their furniture and had it shipped by train to Mountain Park. They sold the house for $80.

Chapter Sixteen
High in the Mountains

◆

Jim and Ida were sitting at their kitchen table for a late evening snack of tea and fresh-made buns with ample butter and jam for Jim. He glanced at the clock on the buffet as the shrill blast of the mine whistle echoed up the valley.

"Two minutes to nine. They're early again tonight."

"Mrs. Avery says it's only announcing eventide," Ida replied. "She told me not to set our clocks by it."

"I'd rather hear the long blast at six that says there's work tomorrow," Jim said briskly.

He didn't attempt to hide his discontent that after a week in Mountain Park, he had yet to work a shift.

"Carl says it will only be a couple of more days until we hear the signal that says, 'work tomorrow'."

The eight-day clock on the buffet began to chime the hour as it had done for them faithfully since they received it as a wedding present from the Davidsons. Ida looked around the room.

"This place is feeling more comfortable every day. We'll do fine here."

With limited housing available, Ida and Jim had been fortunate to sublet the upstairs floor of the company house rented by Mrs. Avery. It was just one big room that they partitioned with a curtain to create a kitchen and a bedroom.

"You've done a good job of setting it up, lass, and it feels more like home now that the furniture has arrived."

"The floors are so warm the baby will be able to crawl around in the winter."

"It'll be Mrs. Avery's coal that heats it."

"They deliver water tomorrow so I'll need to get as much as I can heated for washing before they come."

"I wish we could afford to buy you a washing machine like Mrs. Malloy has," Jim said.

"Well, for one thing, he's the mine engineer and makes more than you do. And for another, they have plumbing. What would we do with a machine up here? Besides you should see how grey her wash is. I wouldn't hang out clothes that looked like that." She thought a moment, then added, "Tomorrow might be a good day for you to try out the McLeod River."

"Maybe I will. Make me a lunch in the morning and I'll see if we can have trout for supper."

The next morning before he left, Jim filled the boiler on the stove, made sure the reservoir was topped up and poured cold water into one of the tubs to be used for rinsing. The almost-empty forty-five-gallon drum on the landing was now ready for refilling.

"Just leave the tubs," Jim said. "I'll empty them while you're cooking the trout."

He set off north along the river until he found a stream pouring into the snaking waterway. The air was fresh on this September morning in 1934 but he found he was breathing a little harder than he usually did when fishing. At 6,200 feet above sea level, Mountain Park was the highest community in Canada and breathing was a bit harder until the body got used to the altitude. He wondered if it would bother him at work.

Experience had taught Jim that trout often congregate in swirling pools where creeks change direction. When he came upon such a spot, he attached one of his colourful flies to the line and began to cast. A few good strikes helped lift his spirits and, as he moved from pool to pool along the stream, he kept adding trout to his creel.

With lightness in his step, he headed home in the late afternoon. As he walked up the hill from the river valley into the town, the dome of Mount Cheviot stood majestically in the distance. Robert Thornton, a British mining engineer who had evaluated the coal claims in the area in 1910 prior to the development of Mountain Park, had named the peak, the mountains reminding him of the Cheviot Hills in the border country of Scotland. As the Elliots had their origins in the Cheviot Hills, for a moment, Jim was walking home from the Devon with his creel of fresh trout to give to his mother.

Ida heard him climbing the outside stairway and went to the door to greet him.

"How did it go?"

The wet creel held the answer, but she wanted to hear it from him.

"There's enough for supper and some left over for Mrs. Avery."

"Do you mind emptying the washtubs before you take your boots off? The water men filled the barrel."

Jim would love to have been able to pour the dirty wash water over the landing but it would make it too mucky for the Averys so, two pails at a time, he repeatedly descended and ascended the ever-lengthening stairway until both tubs were empty.

They were just finishing their tea after a tasty trout supper when they heard the one long blast of the mine whistle. Ida smiled.

"I told you it would get better," she said.

"And none too soon for me."

Jim worked as hard as he always had but it took a few weeks of extreme tiredness for his body to begin to adjust to the difference in altitude. Ida found that it took her longer to do the laundry, but she was not as tired as Jim.

"Maybe if you gave up smoking, it would help," she suggested.

"Dinna be daft, woman. I've smoked since I was eight. I'll get used to it."

Even after they moved to Mountain Park, Ida and Jim continued to have the bulk of their groceries shipped from Woodward's in Edmonton. The trains ran three times a week, arriving late in the day and, as the station was on the other side of the valley across the river from the town, the people would often file down the road and across the bridge to meet the train and pick up their smaller items. It was quite a social event, with the townsfolk filing back up to the village to gather at the post office and wait for the mail to be sorted. Even if the train was late, the postmaster knew that getting mail on train day was a highlight for most people, so he would open up, whatever the time — often after midnight. The village drayman always met the train, loaded the larger orders onto his wagon, and delivered the items door-to-door.

One of the first things Ida did after they settled in was visit the local butcher shop. It was going to be a treat to be able to buy fresh meat after all the canned wild stew they had eaten, and Jim would be able to have the sausages, black pudding, kidney and liver that he had been longing for since she met him. As she asked for her selections, it dawned on her that this was the very first time she had bought meat since she had been married. *I'll never eat wild meat again.*

Ida bought her dresses and some clothes for Jimmy from a woman who periodically came by train from Edson. She was one of many people who set up shop in the local mercantile on train days, having purchased items in Edmonton to sell along the Coal Branch line. The peddlers brought in the latest fashions and the store owner received a commission on all sales. Jim's clothing came from Eaton's catalogue.

About six weeks after their arrival in Mountain Park, Ida woke up one morning with a familiar feeling of nausea.

"Oh, boy," she sighed.

"What is it?" asked Jim, suddenly awake but not sure why.

"I'm not feeling well. I think I might be expecting again."

"Are you sure? I mean, that's good, isn't it?"

"Well, I missed my monthly and it feels like the same awful stuff. I probably am. Could you get me some crackers?"

One thing Ida knew from the outset of this pregnancy was that she would not be running down those stairs and out to the outhouse if she had to vomit. The outside toilet they shared with the Averys had taken some getting used to again. *I hate those dynamite boxes.* She would use the slop bucket upstairs if she needed to. In the early days, this time, she found that some of the advice, like eating dry soda crackers, eased her nausea a bit, and her vomiting was not as violent.

Ida did not attend church in Mountain Park, but some of the people in the community remembered her family from earlier years and soon, the itinerant priest was informed of her presence. Armed with the information his sources had provided, he visited the Elliots one day. Ida invited him in and, even before he sat down, he announced the purpose of his visit.

"It is my understanding that you were not married in the church and are therefore living in sin." He looked past Ida at Jim sitting on the Winnipeg couch across the room and continued, "I think you should do the right thing by this woman and marry her properly in *the* church."

Jim bolted upright, took two giant steps forward with his right forefinger aimed at the priest's chest.

"I'll have you know we were married in a church!" he bellowed. "And I won't have you saying anything bad about my wife or my marriage. So get the hell out of here and don't ever come back!"

Ida did not take her eyes off her husband as the priest silently retreated and slunk down the stairs.

"You were very harsh with him," Ida said.

"Harsh, be damned! He had no right to come here and say that."

"We'll have to think about getting the baby baptized sometime."

"Well, it won't be in the Catholic church! The church I grew up in only baptized adults so there's no rush."

Ida did not believe this. When she took her catechism, she had been taught that all children should be baptized as early as possible, and if they were not, they would end up in limbo if anything happened to them. However, Jim's vehemence determined that this would not be a good time to continue discussing the subject.

In October, as she moved through the early weeks of her pregnancy, Ida was able to be around food more easily than the first time, and so she ordered two cases of pears from Edmonton and decided to try her hand at preserving them. After filling the jars with the boiled mixture of fruit and sugar, she put the lids on, and thinking the process was complete, stored them on a shelf in the kitchen. Within a few weeks, bubbles began to form in the jars and it was obvious that something was wrong.

"Let's open one and see what it tastes like," Jim suggested.

Unscrewing the top, he stuck a knife into the rubber ring that separated the lid from the jar. With a *poof*, a rich pleasant odour filled their nostrils.

"Let's give it a try," he said, lifting the jar to his lips.

"Be careful, it might make you sick."

"Sick, hell, it might make me drunk. This stuff is fermenting. Here, have a taste."

"Ooh, it's kind of like wine," said Ida, her mouth involuntarily puckering.

Taste like wine, it did. There was only one thing to do. Jim collected some beer and pop bottles, picked up some new caps at the general store, borrowed a capper, and returned home to deal with the preserves. After emptying the contents of the jars into the large canner, he gave the mixture a strong stir and he and Ida began to bottle the liquid. When they were done, Jim suggested they store it in a dark place to let it complete its fermenting process. The only space available was under their bed.

As Jimmy grew a little older, Jim and Ida arranged to have someone stay with him whenever they went to a dance at the social club two blocks from their house. The dances were a good distraction from the routine of life in a mining town.

"Mrs. Avery is downstairs if you need anything," Jim told shy Victoria, who was about to embark on her first babysitting job. "I'll come back and check on things during the dance."

"He's been changed and fed and should sleep until we get home," Ida said. "If he does wake up, just rock his basket back and forth and he'll settle down."

When the orchestra took a break and the women began laying out the refreshments, Jim decided to run home to see how things were going. When he opened the door, he heard someone crying and it was not the baby. He quickly pushed the curtain aside and dashed into the bedroom area where Victoria was huddled in a corner, sobbing.

"I want to go home," she cried.

"What's the matter, lass?"

Before she could answer, there was a loud retort from under the bed.

"Something's under there! I want to go home," she shrieked.

Jim bent down to look under the bed, and in the dim light coming through the curtain, he saw liquid running out into the room. His home brew, which had continued to ferment in the bottles, was exploding. He comforted the girl, explained what was happening and asked her to wait in the kitchen until he could get Ida. After he walked Victoria home, Jim and Ida removed the unbroken bottles from the carnage, set them outside on the landing, and then began to clean up the mess. When they had finished, Jim suggested that he might try one of the bottles.

"Don't you dare open that in the house!" Ida said.

He went out onto the landing and, holding the bottle at arm's length applied the opener to the cap. With a *whoosh*, the cap flew out into the street with most of the contents gushing over Jim's hand. Quickly putting the bottle to his mouth, he stemmed the flow with a long drink. Then smacking his lips, he turned to Ida.

"It's no' bad, lassie."

He slept well that night: his stomach warmed, and his mind forgetting how he was feeling about his job. The mine was still not working steadily and Jim was not happy with the way the manager ran the operation. He would not use the contract system that allowed miners to negotiate wages based on output and rewarded the hardest workers. Instead, he set the quotas for the coal cutters because he did not want some men to earn more than others.

It was the first time Jim had worked for a company that did not have Scotsmen in all of the management positions and, as such, he received no special favours.

"I don't think I'm going to stay here long, Ida," Jim said after another discouraging day below ground.

"What are you going to do?"

"We can't move again now that you're expecting, so I've decided to use the idle days to study for my third-class ticket."

"Then you can get a better job . . . somewhere else."

"Well, I'll need it if I am going to become a fire boss. Then I'll get steady work and more pay."

While he was very adept at the practical side of mining, Jim had never studied the terminology that a government certificate demanded. He asked Mr. Farnham, the school principal, to help him with the correspondence course. Soon Farnham was dropping by regularly in the evenings and was a big help to the eager student. However, when he arrived after supper on November first, he appeared upset.

"What's the matter, laddie?" Jim asked.

"Last night. Hallowe'en. And . . ."

"What happened?"

"Some children took the box from my outhouse and set it on my back step then knocked on my door. When I didn't see anyone through the screen, I opened it and stepped right into the mess."

Jim did not know what to say and looked to Ida.

"I guess children don't change much, do they?" she said, struggling to suppress a giggle. "We used to do some terrible things when I was a kid."

"I can't imagine you doing anything as bad as that, Ida."

"Well, I remember one year when we found some paint and spread it around the rim of the seats in all the outhouses." With a mischievous grin, she added, "We never got to see the end result."

"I will probably laugh at this later, but I can still smell it." He looked down at his highly-polished shoes. "I've cleaned and cleaned them."

"I bet they haven't shone like that for years," Jim offered with a chuckle. "Sit down laddie, I'll pour you a wee dram.

Mary and Carl visited often as their children, Margaret and Jean, were now older and could be left alone. In any case, Ida did not like taking little Jimmy to her sister's home. Mary was so fussy that she had put a lock on her bedroom door so children could not get at her things. If Jimmy got fingerprints on anything while they visited, Mary got out a rag and cleaned them up, all the while clucking about sticky fingers. Carl was easy-going and got along well with Jim and Ida, but Jim likewise had trouble with Mary. He did not like the way she treated her children or Ida but especially Carl, who she hen-pecked. Jim did not want to cause any more bad feelings between the sisters than there already was and, since both he and Ida enjoyed playing bridge with them, he overlooked many of her negative comments when they were out together.

Ida continued to write to Doll and Martha, as well as her own folks on a weekly basis, keeping them up-to-date on life in Mountain Park. How good it was to have a doctor in the town for her regular check-ups and how much better she was feeling as her pregnancy progressed. But lately, the letters from Scotland had been very depressing. Shortly after Martha had begun to write, her letters and those from Doll had focused on how difficult life was for Jim's sister, Mary.

Mary had discovered she was pregnant, and when she told her beau, Campbell, he told her that he was married and did not intend to leave his wife. This had left her heartbroken and when her baby, Leonard, was born that summer, she became very dejected and withdrawn. In the spring of 1935, she received a deep cut while working on one of the machines in the mill and the resulting infection did not respond to treatment. Doll wrote in early May to say that they were worried about Mary, who had become despondent and was refusing any medical help for her blood poisoning.

The month of May also saw Ida and Jim preparing for the birth of their second child. The hospital at Mountain Park was small and only set up for mine emergencies, so there were no facilities for delivering babies. At the appointed time, Jim would just have to contact the doctor and nurse and they would come to the house for the birthing.

On Thursday, May 23rd, Ida woke with familiar back pains and suggested that this would be the day. Later, as the pains increased in frequency and intensity, she told Jim to ask Mrs. Avery to come for Jimmy and then get the doctor.

"I'll come right over," the doctor promised when Jim called by his house.

Jim then set off for the nurse, who asked him to wait while she changed from her gardening clothes into her uniform. It seemed to take forever. Finally, she was ready and they hurried along the dusty road, climbed the stairs and were greeted by the doctor.

"You have a beautiful little girl," he announced. "Once the nurse has cleaned up a bit, you can go on through."

The last part of Ida's labour had gone quickly and without as much pain as she had remembered from the first time. Once again, she was relieved to have it all over and later was able to jokingly scold Jim for being late with the nurse.

"I would like to call her Mary," Jim said as they discussed possible names.

"Why does everyone have to have a Mary in their family?"

"Well, I like it and my sister Mary is very special to me."

And your old sweetheart, Ida thought, but she knew better than to bring that up.

"But I have a sister Mary who is not special to me, and certainly not to you," she countered. "I want to name her after Bunty."

"Bunty's not a good name for our baby."

"It's her nickname. Her real name is Catherine."

"Well, that's not so bad. What about Mary for a middle name?"

"I've been thinking of May, the month of her birth."

Ida would not be dissuaded. She did not want to call her new baby Mary. The new baby was named Catherine May and the registration forms filled in and sent off.

Three weeks later, a letter from Doll contained the tragic news that Mary had succumbed to the infection.

"I think she lost her will to live," Doll had written.

Jim was heartbroken. Mary had been his favourite sister, the special wee lass that he had doted on.

"Ida, we've got to name the baby after her. I want to keep her name in the family."

Sitting across the table, Ida took in her husband's anguish.

"We can't do that, Jim," she replied. "She's already registered and everything. We can have another child."

Ida now had to cope with two children under two years of age and was often very tired. One day, while both children were sleeping, her sister Mary dropped by to find her sister resting on the Winnipeg couch. She bustled about and offered her usual bossy suggestions as to what Ida might do to make things neater around the house and insinuated that she shouldn't be tired at her age. As she was descending the outside stairs on her way home, a passing neighbour called out and asked how Ida was doing.

"She's fine," replied Ida's hypochondriac sister. "I'm the one that should be in bed."

When Ida told Jim of Mary's visit and her comments, he was incensed.

"One of these days I am going to tell her just what I think of her," he said angrily. "She is so high and mighty and needs to be brought down. If Carl knew of her shenanigans . . ."

"Jim, you can't say anything to Carl about her. It would destroy him."

"I won't do that. But someone needs to bring her down to size."

Ida continued to use two tubs to do the laundry, but felt she should double-rinse the dark clothes even when she was exhausted by the demands of the children. She washed in one tub and did the first rinse in the other, then, after wringing out the wet clothes, laid them on the oilcloth-covered kitchen table while she emptied the first rinse water and refilled the tub for the final cleansing rinse.

One day, she had just taken the clothes from the tub, placed them on the table and was about to ask Jim to empty the water, when Mary came by again to see how her younger sister was doing. Mary ignored Jim, who was sitting in the corner reading his mining handbook, quickly inspected the room and saw what she deemed to be disarray.

"Look at this house!" she exclaimed. "I'm ashamed to say you are my sister!"

Jim bolted out of his chair.

"You have no right to give anyone advice, woman!" he bellowed. "Get the hell out of here and don't come back!"

Mary left immediately.

"Jim, you really hurt her feelings."

"Someone has to tell her what she does to people. What about your feelings?"

"Well, that's just Mary. I've had a lifetime of learning how to ignore her."

"I haven't, and I won't."

It would be almost a year before Mary was comfortable visiting her sister again.

In June, Jim was successful in his studies, passed his exams in exemplary fashion and achieved his third-class mining certificate.

"Now I can study for my fire boss ticket."

"I'm so proud of you, Jim," Ida beamed. "How about a nice beef and kidney stew for supper?"

That summer, Ida taught herself to knit, starting with socks on four needles. She had picked up a Vogue instruction book and plunged in on her own. She was knitting happily when one of the neighbours, Mrs. Kennedy, came by with a little gift for the baby.

"What are you knitting?" she asked.

"A pair of socks," Ida replied. "I have been teaching myself how to do it."

Mrs. Kennedy looked at Ida's work and gently took the needles from her, examining the knitting.

"Oh lassie," she cried, "you're doing it inside-out. You're going in the wrong direction."

She then demonstrated the correct way to loop the wool over the needles and lift the stitches. Ida's innate left-handedness had emerged in a most creative way. A few days later, she asked Mrs. Kennedy if she could help with the heels as they just weren't working. For a time, Ida knit a number of pairs of pure woolen socks for Jim with Mrs. Kennedy doing the heels. When she mastered that skill, she went on to knit outfits for her children.

In January 1936, one of the miners, Bob Chapman, was crushed and severely injured when caught by falling coal at the face. Because most people did not have much money during the Depression, whenever there was a tragedy, friends and neighbours responded with help as best they could. On this occasion, someone decided that they would raise money for the Chapmans by organizing a benefit hockey game. While each of the communities in the region had teams playing in a league, the organizers of the benefit thought it would be more fun if they made up two teams from miners who were not involved in the regular league.

Jim, who had never had a pair of ice skates on his feet in his entire life, volunteered to participate, which is all that was needed to qualify. As they prepared for the game, they borrowed skates and other equipment and the bumbling players formed up into two equally balanced teams.

Jim quickly mastered the ability to move his feet in short steps that propelled him across the ice at an ever-increasing rate. He was fine as long he kept moving. Unfortunately, the ice surface had a boundary and he quickly discovered that the boards were not the place he wanted to be, at least not when he met them at breakneck speed. Not one to quit, he found he could keep his balance by putting his weight onto the hockey stick and that turning and stopping became much easier if his free hand found another player to latch onto.

By his own account, he was one of the fastest skaters on the ice and actually scored a goal when he got the puck. The other players, not wanting to be dragged to the ice, got out of his way. Fortunately, the puck landed in the net for a legitimate goal just before the flaying Scot followed it in. The goalie, Jack Tennant, who had been assigned to the lamp room after he had lost an eye in an earlier accident, had a problem with depth perception, which forced him to make a number of unorthodox moves to keep players and pucks out of the net, but he had no moves to deal with Jim.

The game was a success financially and raised $600 for the Chapmans. It was also a success in building community spirit, and the organizers decided there should be a second game. Many wanted to see Tennant and Elliot play again. Ida was not one of these. She had been so embarrassed by Jim's outrageous behaviour that she would not attend the second game.

"I'm not going to see you make a fool of yourself again."

"Och, we're just having fun, lass."

"Well, I don't think it's the way a married man with a family should act."

"It's for a good cause. Will you not come?"

"No, but you go ahead. Just don't tell me any stories when you get home."

As Jim came off shift on the day following the second game, he received word that the manager wanted to see him in the office. He did not have the same rapport with this man as he had with previous managers and tried to think if he had done anything that might get him fired. When he got to the office, he was faced with the manager, pit boss, electrician and chief engineer.

"Jim," said the grim-faced manager, "I have an important telegram to read to you."

Jim looked around the room. All eyes were on him, none giving a hint of what the contents might be.

"It's from the New York Rangers," the manager went on. "They want to sign you up to play hockey for them."

With that, the entire room, save for Jim, burst into laughter, but he joined in a few stunned moments later. Even if this seemingly aloof manager did not know how to run a mine, he *did* have a sense of humour.

Jim worked steadily through that winter, but as spring approached, he began to get restless again. Besides being upset by the manager, he did not like the way the foreign single men were being treated in the mine and by the community. They lived in an annex to the hotel and were often excluded from community activities because of their differing customs and lack of English. Jim recognized that most of these men were hard workers and for him that was all that counted. If a man could put out coal, it did not matter what language he spoke. Maybe it was time to move on.

"I'm fed up, Ida. This isn't going the way I thought it would. I don't want to spend another summer here."

Chapter Seventeen
No Relief

"It was worth the wait," Jim announced. "Letters from Craig and Stella."

He waved a couple of envelopes on his return from the post office, just after eight on a Wednesday evening in April 1936. Andy Craig had moved to Nacmine in the Drumheller Valley just after Jim and Ida left Drinnan for Mountain Park.

"I've got the kettle on."

"I'll read Craig's while you make the tea. Craig says there's lots of work and he's doing well." Jim looked up from the letter. "Do you know anything about that part of the country?"

"Not really. I think it is much hotter than the mountains," Ida replied.

The tea steeping, she began reading Stella's letter.

"Stella," she announced, "has talked to Fabbi and he'll give you a job in the dairy."

In her letters to her family, Ida had related how unsettled Jim was in Mountain Park and that he really needed to get out. The mail from Lethbridge always seemed to contain news of Fabbi and the Purity Dairy, especially the letters from Stella who was now working as Simon Fabbi's housekeeper.

"Maybe this is what you've been looking for," Ida said eagerly, as she continued with Stella's letter.

"What's that?" Jim asked.

"The dairy. This is your chance to get out of the mines."

"Well, God knows I didn't do well around cows and I'd rather breathe coal dust than clean a damn smelly chicken coop."

"But there must be lots to do in the dairy without milking cows or looking after chickens." She paused as she reached for the teapot. "Tea?"

"Aye," Jim said.

He held his cup out to her. Jim raised his cup, sipped the brew and uttered his usual gentle "Ahhh," which announced that it was strong enough, hot enough, and sweet enough.

"Lethbridge . . . You'd be close to your family . . . We can't afford to ship our furniture that far . . ."

Ida said nothing. He took another sip of tea.

"We'd have to sell everything and start over."

"I'll do whatever we need to if it will make life better for you."

"Let's hope this is the last time we have to move."

The next two weeks were hectic as letters back and forth to Lethbridge confirmed the job and a place to stay, so Jim and Ida packed their personal belongings, sold the furniture for $60, boarded the slow-moving train to Edson and moved on to their new home in Lethbridge.

While waiting to change trains, Jim decided that the family could afford lunch in an Edson restaurant. Cathy, two weeks shy of her first birthday and all decked out in a new woolen dress and matching soakers that Ida had knitted, was comfortably seated on her mother's knee when the waitress brought coffee to the table. She set Ida's cup right in front of the baby, who immediately reached for it and spilled the scalding contents onto her lap and into the absorbing woolen garments. The resulting scream brought the waitress rushing back to the table, followed by the owner, who hustled the family into a nearby room where Ida completed the removal of the steaming garments from her howling daughter. Ida quickly sponged her with cool water. The skin was a reddish colour but no blisters had formed and the child was soon calm enough for them to return and finish the meal. Jim was furious with the waitress but had no opportunity to express his feelings toward her as the apologetic owner attended their table until the meal was completed. He waived the bill for the lunch.

Simon Fabbi met the family at the station and drove them to their new home, a cabin behind the rooming house owned by Mrs. Berti, a friend of Angelina. With three furnished rooms, it was a bargain at $10 a month and just a short walk to the dairy.

When Simon bought the dairy, his sons designed a logo for the milk bottles that included the motto — *Yours to love, ours to protect* — and a silhouette of Romeo and Beulah's son, Gary, painted on each bottle.

Simon had been successful with hard work and frugality, and he believed that everyone in his employ should work as hard and as long as he did and do whatever job he decided was pressing at that moment.

Jim had been used to working regular hours and getting paid according to his production but the dairy was not at all like that. The small monthly salary reminded him of his farm days when there was no relationship between effort put out or hours worked and pay received. Frank and young Johnny Linteris were content with the arrangement at the dairy, but Pete had quit and got a truck-driving job. Jim, however, was troubled by the work situation and told Ida that he had made a mistake in moving to Lethbridge.

"Try it for a bit longer," she said. "Maybe things will get better."

"Well, they'd have to improve a lot."

Then one evening, Pete came over for tea and told Jim and Ida some of the things that had not been included in the family letters.

When Frank and Angelina had returned to Lethbridge, they had settled in the river bottom area under the High Level Bridge where housing was less expensive than other parts of the city. Johnny and Pete, who worked with their father at the dairy, boarded in Mrs. Berti's rooming house and only visited their parents on the occasional weekend.

One Sunday when they arrived for supper, their mother had not greeted them at the door but busied herself over the stove as they entered the house.

"Come, let's have some wine while your mother cooks," Frank said. "She is such a bad cook, she has to put all of her mind into what she is doing."

"Mama's a good cook! What are you making?" Pete asked. "It smells good."

"Spaghetti with sausage," she replied quietly.

"Are you going to take all night, woman?" Frank barked at Angelina.

When she brought the bowls of food to the table and sat down, she could not hide the bruises on her face.

"What's this?" Pete demanded.

"I slapped her," said Frank. "Keeps her in line."

"Slapped her, you rotten bastard?" Pete raged, bolting up and grabbing his father's arm. "You *punched* her!"

Johnny jumped up, too, and the two boys wrestled their father out of the house and down the path toward the river. When Frank kicked out at Pete, all three fell to the ground, but the boys would not let go and dragged their father to his feet as they reached the river.

"You're going in, old man!" Pete hollered.

"Wait!" Frank pleaded, digging his heels into the clay. "I won't hit her again."

"You've said that before," Pete threw back. "C'mon, Johnny, in he goes."

His footing gave way and once more the three were in a heap on the soft bank.

"I promise on my mother's grave not to hit her again," Frank yelled.

"If we let you up and you ever hit her again, I will kill you," Pete told him between deep breaths. "Do you understand?"

Frank nodded.

"Yes, yes . . ."

"I think he means it," Johnny said.

The boys slowly released their grip and the three walked back to the distraught Angelina. The family ate in silence — Pete's glances at his father reinforcing his words at the riverbank.

"Too bad you didn't drown the bugger," Jim said.

"Well, I'll tell you that we were close to it."

"If he ever touches her again, Pete, just let me know and I will take care of it."

"Don't worry, Jim, I think we scared him enough that he will keep his hands to himself. But if he doesn't, I'll settle with him."

In the late summer, Simon Fabbi took Frank and Jim out to lunch one day to talk about Stella.

"I care very much for your daughter, Frank, and I want her to be with me," Simon began.

"She's quite young," Frank mumbled.

"That doesn't matter to me. We get on well together and I want to take her on a trip to America."

"She's never been far away from home before."

"You know me, Frank, I will take good care of her. I would like her to live with me."

"Angelina would not like that," Frank said weakly.

"I will make it worthwhile for you and Angelina."

As Jim listened, he realized that Frank was just going through the motions of presenting vague objections but not really meaning any of it.

"Fabbi," he said abruptly, "are you and your wife divorced yet?"

"That's being looked after," Simon replied. "Everything will work out fine."

"Fabbi has been good to our family, Jim," Frank offered, nodding at Simon.

"The two of you dirty old men should be ashamed of yourselves!" Jim hollered as he stood up. "Look at you Frank, making a deal for your daughter. And you, Fabbi, buying your way through life. What kind of people are you?"

With that, he left and went back to work. The two men sat in silence for a moment, until Simon Fabbi raised his glass.

"To a happy future for all of us."

Frank smiled, clinked his glass on Simon's and took a full swig.

That evening at supper, Jim was fuming.

"What's happened now?" Ida asked.

"It's Stella. The way your father and Fabbi talk about her. Like she is something they can trade."

"What do you mean?"

"It's almost like the old days of dowries. Nothing specific was said, but I think Fabbi is giving your father money in exchange *for Stella's hand*."

"Oh, I hope not. But Father is a different person around Fabbi and agrees with whatever the man says."

"I think of how badly my sister Mary got hurt with that gink Campbell, and what that did to her. I don't want Stella to go through anything like that."

"She's almost twenty-one, Jim, and as much as we might not like it, Stella will do whatever she wants."

"You're right," he replied with some resignation. "But I'll kill the bugger if he hurts her!"

Jim continued to struggle with the work at the dairy and his increasingly negative feelings toward Frank. The hours were long, the pay low, the summer hot and the smells offensive. At least in the mine, he could work in the cool, the days were predictable and the wages were good. Then one hot September afternoon, Simon decided they should re-tar the roof of the dairy. As Jim was trying to get the last of the hot tar out of a pail and onto the roof, his grip slipped on the paddle and the hot liquid splashed up over his arm and onto his face.

"I'm scalded," he hollered, excruciating pain driving into his body.

Frank mumbled something as another worker helped Jim down the ladder where Fabbi poured a bucket of cold water over him and cooled the black mass. Then it was into Fabbi's car for a quick trip to the hospital where the emergency room staff carefully removed the tar. Since he had not been burned deeply enough to produce blisters, the nurse had smeared a salve on his face and arm and discharged him.

On his return home, Ida immediately noticed his red oily face and his tar-spattered clothes.

"What happened?" she exclaimed.

"Och, it wasn't much. Spilled some tar on myself. It hurt a lot at first, but I'm fine."

"It looks sore."

"You know what your father said as I was yelling for help? 'It's your own fault for being so careless.' Ida, I have to get away from him and the dairy."

Ida had not experienced the kind of the frustration Jim had during that summer. It had been good for her to have Nonna and Nonno — grandma and grandpa — so close to the children and Stella in her life again. Lethbridge was familiar to Ida and she was developing a good friendship with Romeo's wife, Beulah, and with Stan's wife, Antoinette. She and Jim also played bridge frequently with the Friegrans, a couple who had moved to Lethbridge from Drinnan the year before. Life was much easier for her here even if they did not have much money. She was not sure she wanted to move away this soon.

"What are you thinking?" she asked.

"Ida, we have no money. The measly wage Fabbi pays barely keeps us even. I think I'll write to Craig and see how the work is there."

"What about one of the mines here? Carl Friegran seems to be doing okay."

"No. I have to get away from your family."

Andy Craig replied by return mail, with the news that he had talked to the pit boss at the Commander mine, at Nacmine in the Drumheller Valley, who assured him that he would hire Jim as soon as he arrived. The letter also contained a ten-dollar bill to help Jim get there. "You can stay with me until you get settled," Andy concluded the letter. That was enough for Jim and once more plans were underway for a move.

When she heard the news, Mrs. Berti offered to let Ida and the children stay on until Jim found a place in Nacmine for the family. Early the following morning, Jim packed a few things into his duffel bag, took two dollars to spend on the way and set out to hitchhike to Nacmine. He was picked up by some very generous drivers along the way and in just four hours travelled the 135 miles to Calgary. There he took a transit bus across the city to the eastern boundary and the start of number 9 Highway. In a few moments, a man who owned a second-hand store in Drumheller picked him up.

"Where are you headed?" the man asked.

"Nacmine. Do you ken where that is?"

"I'm on my way home to Drumheller. It's about four miles up the river valley from there. Where you from?"

"Scotland. I'm heading to a new job at the Commander mine."

The ninety miles to Drumheller took almost three hours and in that time, Jim told some of his story and the driver responded to his honesty.

"When you get settled in, come and see me and I'll help out with furniture."

"Och, I've no money and it will take a while to get on my feet."

"Just come to my store and pick out what you need. Pay me later. I think you should get your family here as soon as you can."

One more ride took Jim to Nacmine, where he looked up Andy. The next day the pit boss at the Commander mine hired Jim and he began house-hunting.

At the end of that month, Ida did not have the necessary money to pay the rent, and she asked Mrs. Berti if she would mind waiting until Jim was paid at his new job.

"I know your family and I am not worried," Mrs. Berti said. "But if you go on relief, they will give you money for rent and food."

"Jim would never let me go on relief. He feels very strongly about that. Please let us stay."

"There is no problem, Ida. I just think you would find it easier if you had some help."

Ida did not wish to go against Jim's wishes so she managed with free milk from the dairy, vegetables from her garden and baking from her mother and Mrs. Berti. But just two weeks later, Angelina and Frank arrived at the door on a Saturday afternoon, looking distraught and dishevelled.

"Mama, what's the matter?" Ida asked.

"The house," Frank said sharply. "It burned to the ground. Everything is gone."

Ida put her arm around her mother.

"How did it happen?" she asked.

"We were up the river getting wood and when we came back it was pouring out smoke."

"The fire department?" Ida queried.

"By the time they came, all they did was cool the ashes. We've lost everything."

"What will you do?"

"We have nowhere to go," replied Frank. "We'll have to stay with you."

They moved in and, even though it was crowded, Ida, with her hard-working mother helping, got some respite from the continuous washing and childcare. About a week after the fire, Frank announced that he'd had enough of Lethbridge and, with winter coming, had decided that they should move to Vancouver. It puzzled Ida how quickly he made the decision and how he could afford it. Maybe Jim's suspicions had been right.

It took another two months before Jim could put aside enough to pay rent and send the train fare to Ida. In the meantime, he had found a place to rent for his family — one of six two-room shacks built by an American speculator along the bank of the Red Deer River, on the eastern side of Nacmine and about a mile from the Commander. Once the details had been worked out, Jim walked into Drumheller to visit the second-hand furniture dealer.

"Hi, laddie," Jim beamed as he entered the store. "Do you remember me?"

"Scotland. I sure do. How's it going?"

"I've a job and a place to stay. Now I need furniture."

"Pick out what you like and I'll have it delivered."

"But I've no money yet."

"I told you before. Pay me when your family arrives and you get on your feet."

"You'll no' regret this, laddie."

When Ida arrived, she looked around the simply furnished, tiny house. *At least it's clean*, she thought. Jim proudly lifted the trapdoor in the kitchen floor to show her the dugout cellar and the box on the stairs with milk, eggs and butter in it.

"This will do as well as an icebox, Ida."

The furniture Jim had chosen was old, but Ida had brought bedding and towels with her and she was soon able to create a homey feeling.

It was hotter in this arid valley than it had been in Lethbridge. Fortunately, the willows and cottonwoods along the riverbank provided a cool spot to sit on a hot afternoon. A dry creek bed ran along the edge of the yard, a reminder of pre-dustbowl days.

A few weeks after the family had settled in, Jim came home with a dog: a mid-sized creature whose lineage was obviously pure mining-town blend.

"One of Andy's friends gave it to me. He'll be good for the wee ones. I think we should call him Rover."

Jimmy and Cathy took immediately to the tail-wagger and he to them and, without any training, he assumed a watchdog role with the children and provided a noisy barrier between them and the river.

The Commander mine worked from September 1936 to February 1937, and then, since enough coal had been stockpiled for the railroads and local consumption, it shut down again until fall. During that working period, Jim and Ida had been able to send the final rent money to Mrs. Berti and pay off the furniture dealer, but when the layoff came, they were broke.

"What are we going to do?" Ida asked as they sat looking at what would be Jim's last pay for about six months. "I'll plant a garden in the front yard as soon as it gets warmer, but we have to do something now."

"I know, hen," Jim replied dejectedly. "There seems to be only one way out . . ." He paused, before forcing himself to say the word. ". . . relief."

"It won't be so bad, Jim. Lots of people do it."

"I left Scotland to get away from being on the dole and here I am."

The next day, he walked the three miles to Drumheller and entered the RCMP office where relief applications were handled. He was told that his completed forms would be sent to Edmonton for approval and he would hear later if his request was successful or not.

Over a week later, a letter arrived from Edmonton stating that, if Jim had saved his money when he was working, he would not be in this unfortunate position. The next day another letter — this time from the Mounties — told them that, if they went on relief, the RCMP would have the right to visit the family at any time, and if they found even a small amount of extra food stored in the house, they would be disqualified from any further relief.

"Bugger this!" Jim said.

He put on his jacket, stormed out of the house and set off for Drumheller. He stomped into the RCMP station, waving the letter.

"Are you telling me," he demanded, "that if we get relief, we can't have so much as a bag of flour or vegetables in the root house?"

"Yes." A firm and blunt reply.

"You're saying that, with two children needing to be fed, we are not allowed to have any spare food at all?"

"That's what I said. Don't you understand plain English?" the officer replied briskly.

"You're damn tootin', I understand. You have no idea what it takes to raise a family."

No Relief 217

Jim threw the letter on the counter.

"Take your bloody relief and give it to someone who will kowtow to your stupid regulations! If I had a 'ski' at the end of my name, you'd give it to me."

He stormed furiously out into the street and stood for a few moments to catch his breath. What now?

He took out one of his roll-your-own cigarettes, flicked the wheel on his bullet lighter and inhaled the soothing smoke. *Will this never end?*

Looking at his reflection in a store window, he straightened his jacket and adjusted his cap. Taking a deep breath, he faced his next challenge. He went to Superior Meat Market and then Cruickshank Groceries and asked them, in turn, if they would carry his account until the mine started working again. Both stores delivered to the mining towns three times a week and were familiar with their regular customers and the pattern of layoffs at the mine, so they agreed to allow some miners to run charge accounts all summer. Jim was recognized in both businesses and with a handshake, his family's groceries and meats were looked after until the mine started up again and he could begin paying off the debt.

Cathy liked to be outside, and she was often out playing in the sunshine about the time the young man from Superior Meats came by with his delivery. As he knocked at the door, he would shyly look sideways.

"Mrs. Elliot," he would report, "your daughter is in the yard with no clothes on again."

It seems that when nature called, Cathy would enter the outhouse, take off her clothes, tend to her needs and then exit with clothes either on the seat or down the hole, depending on how careful she was.

"Thanks, Billy," Ida would reply with a smile. "I'll tend to her."

The tending often meant taking the rake with her to retrieve the soiled clothes from their smelly resting place. The outhouse was not a pleasant place to clean and, having a young son who was careless with his aim, she was always scrubbing the seat. In the winter, this meant pouring boiling water on it to break down the mound of ice that would grow from too many dribbles.

The clay banks of the wide Drumheller Valley rose sharply to the fields above and provided a great reflector for the hot summer sun. Not only was the economy depressed at this time, but the terrain showed the stress of little water and the relentless sun. In the evenings when heat rose from the baked clay banks, the air along the river was displaced by colder air from the west and soon a wild wind would develop and whip the stinging dust into a frenzy. Many an evening, the family would look out at the swirling dust and tumbling Russian thistles rolling past their windows or watch the sand slowly sift under the door onto the linoleum. How different this was from the mountains of the Coal Branch.

However, during their first spring in Nacmine, they had discovered that the river lowlands provided an ideal site for growing a vegetable garden. Johnny came from Lethbridge to help Jim spade the yard while Ida planted seeds. Their proximity to the river provided the shallow well with plenty of water, so Ida kept her garden vital during the harsh summer that followed. Jim often took the children down to the river for a swim to cool off. He had learned how to swim in the Devon and had a style that saw him moving through the flowing water with his head high. He would put Jimmy on his back and swim out to a small island with the boy hanging on tight. Cathy was left in the safekeeping of Rover, who would not let the children anywhere near the water when Jim was not around. Jim would then return for Cathy after issuing strong warnings to Jimmy about not moving anywhere near the edge of the water while he was gone.

Ida and Jim were not going to let the Depression destroy them, and they maintained a positive attitude through it all. Ida continued to make bread, which was great for filling them up, and that, coupled with an abundance of eggs from a neighbour's hens, supplemented the groceries prudently purchased. A frequent Saturday night treat for the family was a walk into Drumheller with Jim carrying Cathy on his shoulders and four-year-old Jimmy hopping from tie to tie as they followed the tracks into the city. They would stroll through the small downtown area, stop at Ozzie's for five-cent ice cream cones for the children and then walk home. There was never enough money for the parents to have a cone. When the carnival came, there was no money to spend on it, but it was a time to walk around and see others and so they made the best of it.

The main entertainment for Jim and Ida during this workless period was playing bridge with the next-door neighbours, Monace and Mollie, a childless couple who came over almost nightly. But as those evenings grew longer, Jim began to wonder what he could do to encourage the neighbours to leave earlier.

"It's fine for them to stay up late, they don't have children getting them up early every day," he said over coffee one morning.

"I don't want to hurt their feelings by asking them to leave," Ida replied.

"Well, I'll think of something."

That evening, before Monace and Mollie arrived, Jim took a package of chocolate Ex-Lax laxative, broke it into pieces and put them in a small dish on the table. As they played cards, he quietly monitored the intake of the squares. When he thought that their visitors had consumed a sufficient amount, he picked up the dish and put it on the cupboard.

"I think I'll save a few for the bairns," he said.

The next morning, Jim watched for the neighbours and when he saw Monace head for the outhouse, he went into the backyard and waited. Soon Mollie came rushing out of the house and, seeing Jim, hollered over to him as she waited for Monace.

"Are you all right? Monace and I have had really upset stomachs this morning."

"I'm fine. What's the problem?"

"We were wondering if there might have been something wrong with the salmon sandwiches Ida served last night."

She traded places with Monace, who also asked Jim if he and Ida were okay.

"Aye, we're both fine. Maybe you've got a touch of flu."

Monace and Mollie did not stay as late that night.

After supper one stifling hot July evening, Ida was in the garden when the deep roll of distant thunder drew her eyes to the west. The dark blue clouds with grey swirls and the telltale streaks reaching down to the ground were being split by lightning. As she watched, the sky became darker and the time between the flashes and the increasingly loud thunder grew shorter and shorter. Jim and the children came outside in response to the sound.

"We're in for it," she announced. "Those grey streaks look like hail. We'd better put some boxes over the tomatoes."

Just as she and Jim finished, the sky turned eerily black and the first large drops of rain began to splat against the house and onto the garden. Scurrying inside, they watched the increasing torrent flatten plants and begin making puddles everywhere. The lightning was brighter and the simultaneous thunder could be felt as well as heard. Then, like a passing freight, the noise soon lessened and the lightning diminished, although the deluge kept up, continuing relentlessly throughout the night.

When they woke in the morning, Ida lifted the trapdoor to the cellar to get some milk for breakfast and found the butter floating in water just below the floorboards. She ran to the kitchen window and gasped.

"Jim," she called out, "look how close the river is! It's almost up to the toilet. The yard is full of muck."

They dashed to the front door and saw that their promising, carefully-tended garden was now an ugly mass of mud.

"Oh my God," cried Ida. "It's all gone."

Rover rushed out, slopped his way into the garden and began digging into the mess.

"Get out of there, you stupid dog!" Ida ordered.

Rover was not deterred.

"He's found something," said Jim.

He put on his gumboots and went out to inspect the damage. As he got closer, he saw that the dog was in the process of uncovering a horse's hoof. Jim grabbed a shovel from behind the house and began to dig. It didn't take long to discover the hoof was attached to a leg, which was attached to the rest of the body. Somehow, the flooded creek had washed the horse from one of the farms above the valley and it now lay buried along with Ida's garden.

When Jim inspected the outhouse, he felt it was too close to the swollen river and was in danger of being washed away.

"Ida, we can't stay in this place another day."

"What will we do?"

"I don't know, but start packing our clothes. I'll see if anyone at the mine office knows of a place to rent."

Within three hours, he learned that Art Hill, the mine accountant, had a house for rent and that they could use the coal truck to move there — which they did that same day. The new location was at the opposite end of the community from where they had been living and was closer to the mine, Ayling's store, the post office and the mine executives' homes.

Soon after they settled in, they began to develop new friendships, especially with Jean and John Henderson, who had two children, Jean and Ian, about the age of Jimmy and Cathy. John was the mine electrician and found in Jim someone who understood a great deal about how a mine works. His mother — Granny, as they called her — had lived with them since the early years of his marriage and was an integral part of their family life. She soon became Granny to the Elliot children too, as she and her grandchildren would spend every Friday with Ida. Jean and John went shopping during this time, taking advantage of having time away from his mother for the first time in their marriage.

The cycle of work during the winter when bills were paid followed by a summer of charging groceries again continued through 1938 and into 1939. Jim and Ida found that they had just one paycheque free of paying grocery bills before the mine shut down again for the next season. Jim continued to smoke but saved his butts so that he could open them and re-roll the tobacco into cigarettes.

Shortly after the mine shut down in early March, Ida discovered she was pregnant again.

"Good show, lassie," Jim responded to the news. "I hope it will be our Mary."

Ida smiled. *I pray to God it's a boy*, she thought to herself.

Once again, her bouts of nausea were not as severe as they had been during her first pregnancy.

In May 1939, in spite of the shortness of funds, Jim thought it was important for the family to travel the ninety miles to Calgary to see King George VI and Queen Elizabeth as they passed through on their Canadian tour. Jim, a patriotic Canadian and proud to be part of the British Empire, wanted his family to see royalty. They found a spot right on the curb on the main street, 8th Avenue, and had a perfect view of the approaching royal couple. When their car was right in front of him, Jim let out a loud "Heuch!" — the cry of the dancers at a Scottish ceilidh. Queen Elizabeth immediately turned and, seeing the source of the greeting, waved.

"Did you see that?" Jimmy exclaimed. "The Queen smiled at my daddy!"

That fall, before Jimmy was to start grade one, Jim decided to change jobs and got on at the Monarch mine near Kneehill Crossing two miles along the rail line toward Drumheller. This meant another move for the family and they found a house to rent in the flats below the mine. But on September 10, Mackenzie King declared war on Germany.

"Ida," Jim announced on hearing the news, "I want to join up and serve our country."

Her reply was firm and final.

"You are not going to go gallivanting off to fight some war and leave me with two children while I am six months pregnant."

Chapter Eighteen
War Clouds

The war in Europe was not the only conflict that started in September 1939. Hitler's action in Poland triggered a conflict in the little house on the riverbank below the Commander mine in Nacmine, one that would not be quickly or quietly ended.

As the daily radio reports painted a picture of the rapid advancement of the Nazis in Europe, Jim began a relentless campaign of his own.

"Ida, I have to join up. That gink Hitler is too close to home for me."

"There is no way you are going into the army."

"But all of my family is over there and Britain could be next."

"Not *all* of your family!"

One afternoon, after receiving a letter from her father, Jim was incensed.

"I don't know what's wrong with you, Ida, everyone's joining up. Look! Your brother Pete was the fifth one to sign up in Edmonton. It's what men do."

"Not when they have a family. Not when they're married to me," Ida replied emphatically.

"It's my duty to help protect people from the likes of Hitler. If he has his way, he'll control the world."

"Someone else can do the fighting. You won't be part of it if I have *my* way."

Almost seven years of marriage to Ida had taught Jim that she was no pushover. He realized that he would not come out the winner in the best of times, let alone when Ida was six months into another nauseous pregnancy. He would not press this point further yet.

When their second daughter arrived on December 19th, 1939, they were both pleased that the ordeal was over.

"She'll be Mary," Jim affirmed.

"You know I don't like that name," Ida quickly answered. "We've talked of this before; when my sister died, you said we would name the next girl after her," Jim said solemnly.

"I know, but I don't like it by itself. What do you think about Mary Ann?

"As long as Mary's there, it'll do fine. Ann's a good second name."

"No, I mean Mary Ann as her first name."

"Mary Ann," Jim repeated slowly.

"Yes, and Margaret as her middle name. It sounds good. Mary Ann Margaret Elliot. What about it?"

Jim ran the name through his mind a couple of times, and thought of the second daughter in the royal family.

"She'll be our princess," he said with a chuckle, "my bonnie wee lassie."

With the mines working steadily that winter, Jim and his friend Andy Craig decided that they would take night school courses to work toward their fire boss tickets. Andy had a car, so the two of them could ride to class together each week.

Following his first class, Jim and Ida sat at the kitchen table over tea and ginger snaps.

"Just what does a fire boss have to do?" she asked.

"Well, I'll be the one who fires the shots at the coal face."

"What do you mean?"

Ida suddenly realized that, although her father and her husband were miners, she did not know a great deal of what went on underground.

"I will have to make sure the holes are drilled properly into the coal face, and then put in the right amount of explosives into the hole, clear the area of miners and then set it off."

"I didn't realize it would be so dangerous."

"Ida, it is not dangerous if you follow the rules. The fire boss is the one who makes sure everything is safe before the workers can return and begin removing the coal."

"That's a lot of responsibility."

"I ken that, lassie. That is why I am also taking first-aid training, because I will be in charge of safety for about seventy men and the horses in my district underground. The fire boss examines the gas levels and all the tunnels, airways and timbers for proper safety. Then, during the shift, he twice re-examines his district for safety. That's why my pay will be better."

Andy's childhood sweetheart, Meg, had finally come to live with him and he was overjoyed. She had left Scotland and gone to New York to live the high life, while Andy had stayed in Alloa. Later, he had followed her, only to discover that she had married an Italian and had a little girl, Betty. Andy was heartbroken, and he had made his way to Drinnan where the Alloa Coal Company had started a new mine; it was there he had met Jim and their friendship began. In the meantime, life had not gone well for Meg. She and her husband had parted ways and she was left destitute. But Andy had kept in touch with her and, when she wrote to tell him that she had nowhere to go, he sent her money and invited her to come and live with him. She accepted and had arrived with Betty in the summer of 1939.

Andy had little experience as a husband and father and relied on Jim and Ida to be supportive of him and Meg as she settled in to her new life in a mining town. So each week, Andy dropped Meg off to spend the evening with Ida, while he and Jim attended night school. It worked out fairly well, except that Meg had grown dependent on the use of alcohol and Ida was not comfortable around excessive drinking. She would have the odd glass of wine but was not used to entertaining with drinks.

One cold winter's night, as the two women sat beside the glowing pot-bellied heater, Meg asked Ida if she had anything to drink.

"Well, I have some beet wine out in the shed."

As a child, Ida had watched her father make wine and she and Jim had experimented while in Mountain Park, so with a successful garden that summer, she had decided to try her hand at beet wine. Her methods were primitive as was her bottling style. The product was stored in glass preserving jars in the unheated shed behind the house.

"Could I have some?" Meg asked.

"Sure, just a minute."

Ida threw her coat over her shoulders, slipped into her snow boots and exited the back door. She soon returned with a couple of quarts.

"It's frozen solid," she said.

"That's fine, lass," Meg said quickly. "We'll just set the jars in the warming oven and they'll soon be ready to drink."

As the women continued talking, Meg got up periodically to see how the wine was doing. Each time, she noticed that some thawing had taken place. Soon, she opened both jars and poured off the available liquid into a glass.

"Would you like some?" she asked Ida.

"No thanks. I'm nursing the baby and I don't think it is good for her."

So Meg drank, waited, poured, and drank again, not realizing that almost pure alcohol was the first to thaw. When the men returned, her behaviour was quite obnoxious as she slurred sexual innuendos at both of them.

Jim had struggled with Meg's drinking from the outset and did not like what it was doing to his friend. As soon as the Craigs left that night, he turned to Ida.

"That woman will not get another drink here," he said.

He brought in all of the wine and when it had thawed, poured it into the snow.

The next morning, Meg sent Betty over to Ida and Jim's. They had become quite attached to the bright young girl and worried about what life must be like for her with her mother's inability to control her drinking. Jim jumped to the door the instant he heard her knock.

"Come in, lassie," he said.

She stamped the snow off her feet and quickly shut the door behind her to keep the cold out.

"My mom's not feeling well and wonders if you have any of that wine that might help her."

"You tell your mother if she wants any of that," Jim said as gently as he could," she will have to come over and eat the snow."

"I don't understand," Betty said.

"Ida told me the wine was frozen," Jim offered. "I thought it might be bad, so I have thrown it out."

When Ida returned from the post office one afternoon, she handed Jim a letter from Doll. After he had finished reading about food shortages and rationing and the fear of German submarines off the coast, Jim looked intently at his wife.

"Ida, I really ought to be over there. My family is in danger."

"Not on your life!" she responded.

"It's my duty. Britain is all alone in this."

"It's your duty to look after your family here."

"You'll be looked after. Other wives are doing it," Jim said gruffly.

"I'm not other wives. You can do your part right here digging coal. The mines are working full-time and they need all the coal that can be produced. This is where you belong!"

Her tone of voice said *end of discussion*.

That summer, when the Germans began to bomb England, Jim brought up the subject again.

"Ida, I am a British subject and I need to be there."

"It's bad enough having Pete over there without you going, too. War is awful."

"That's why I want to help stop it."

"It will stop without you!"

One morning in June 1940, shortly after Canada had declared war on Italy, Ida was on her way to the post office when she ran into Mrs. Walsh, a middle-aged woman from Scotland, who asked Ida how she felt about Canada being at war with Italy.

"I haven't thought too much about it, to tell the truth," Ida responded.

"Don't you feel rather badly?" Mrs. Walsh pressed.

The insinuation was that Ida ought to feel some guilt, given that she was Italian. With great pride, Ida informed Mrs. Walsh that her brother Pete was the fifth person to register for the armed services in Edmonton and that he was in North Africa and might even have to shoot his own cousins.

"And," she continued, "unlike you, Mrs. Walsh, I was born in Canada! So why don't you just go back to Scotland with your stupid ideas?"

With that, she turned her back and continued on her way.

On her way home, she dropped in to have tea with Jean and Granny Henderson and told them of her encounter with Mrs. Walsh.

"Good for you, lassie," said Granny with a grin. "That old biddy has too much to say most of the time."

Later that summer, the RCMP arrived at the home of Carl and Mary Franceschi in Mountain Park.

"Mr. Franceschi," the officer stated. "You are considered an alien by the Canadian government and I must confiscate any firearms that you possess."

"I am a naturalized Canadian. I took out papers!" Carl protested.

"You came from Italy and that makes you an alien. Do you have any guns?"

"Just my hunting rifle."

"You won't be able to keep it."

"But that's how we get our meat!"

"You have thirty days to give it away or sell it. And I must warn you that we will be keeping an eye on you and the other aliens in this town."

The officer turned and left.

"I came to Canada to get away from this," Carl said to Mary. "There goes my hunting."

When Carl told his mining partner what the police had said, the man said he would look after the rifle until the war was over. He also agreed to take Carl hunting and, if the opportunity arose, Carl just might *borrow* the rifle for a shot.

Mary wrote to Ida to tell her what was happening.

War Clouds

"He is devastated. He lived for his hunting and now he just won't leave the house."

"He is one of the gentlest men I have ever met," Ida said, telling Jim of the letter.

"They're really giving anyone of German or Italian roots a bad time," Jim offered. "They think everyone's a possible enemy. They should be concentrating on the really bad ones."

They both knew that many Italians were being rounded up in the big cities. For Ida, this just didn't make sense. But then nothing about war made any sense to her.

As the mines were working steadily, for the first time, Jim and Ida were able to pay off their bills in the winter and save a few dollars. They even bought their first washing machine when the Elliot name came to the top of the waiting list. That first time after she had filled the washer with hot water, added the soap and put the diapers into the churning liquid, Ida stepped back and marvelled that her clothes were being cleaned without any back-breaking scrubbing. She set about doing other chores while the wash was being done and noticed that the steady hum of the motor and the sloshing of the clothes had a soothing effect on six-month-old Mary Ann as she lay in her crib.

In the spring of 1941, Jim and Ida decided to try their hand at raising chickens and ordered 200 leghorn pullet chicks. They arrived in a cardboard box with round breathing holes cut out of the side. Jim made a space for them on the floor beside the kitchen stove, where they stayed until they were strong enough to handle the outdoor temperature in the crude coop he had built. The chicks grew quickly as the weather warmed and would provide many a meal later in the year.

But another flood hit the area that summer and, once again, it was time to move to higher ground. Andy had told them about a call for bids on a house at Kneehill Crossing, near where he lived. This was a community of just ten houses, was located at a wye on the rail line between Drumheller and Nacmine, where a spur line crossed the river on an iron bridge to service Midland. The bank had repossessed the house when the owner, a Russian immigrant, was unable to keep up payments, having been committed to the mental hospital in Ponoka. After looking at the place, Jim and Ida decided that they would bid on it. Jim chatted up the agent in charge and told him that all he could afford was $60.

"We'll see what we can do," the man said. "My friend Andy Craig has put in a good word for you."

"Aye, we go back to Scotland," Jim replied. In fact, they had only met in Drinnan.

A week later, the man came by to exchange the deed to the property for three crisp twenty-dollar bills, and the family moved one more time. It was the largest house they had ever lived in — two bedrooms, a living room and kitchen, with a summer kitchen in the back beside the garage. The only drawbacks were the calcimined walls and ceilings, every inch of which had been covered by cobalt-blue calcimine, a mix-ture of zinc oxide, water, glue and colour that dries to a dull, plaster-like finish.

"It's ugly," Ida declared. "The first thing we need to do is paint."

"We can do that," Jim offered.

However, paint proved to be too expensive and calcimine came in many colours. Ida chose light yellow for the kitchen and a pale rose for the bedrooms. Jim borrowed a ladder from Andy and set about painting their new home; it took three coats to obscure the original blue.

The family settled in quickly and Rover found a new role in keeping Catherine in the yard and away from the road. Meanwhile, Jim took advantage of Andy's borrowed ladder and climbed up into the attic through the trapdoor in the kitchen ceiling.

"Ida, look what I found!" he exclaimed.

He hurriedly descended the ladder, waving fists full of pink, yellow and blue paper.

"What have you got?" Ida asked.

"Money! But it's not English. There's a suitcase of it up there along with a bunch of clothes. Must have belonged to that crazy Russian."

A few days later, Jim took some of the money to the bank manager in Drumheller.

"I've come upon a fortune!" he said eagerly.

He pulled a bundle from his pocket.

"Looks foreign. Let's have a look."

"There are some pretty big bills there," Jim continued.

"You're right. It's from Russia, but let's have a look at the dates on them."

He began to flip through the bills.

"Does the date matter?"

"Yes, very much so with Russian money. See, these are all between 1898 and 1915."

"What's that got to do with it?"

"The Revolution. After the revolution in 1917, they printed new money and everything that had been connected to the Tsar was made worthless."

"You mean it's no good?"

"I'm afraid that is just what I mean. Someone may want a souvenir, but this money has no cash value."

The road home seemed longer for Jim that day. He was glum as he told Ida of his disappointing encounter with the bank manager.

"It must have been a terrible letdown for the man who owned this house to bring all that money to Canada and find it was no good," Ida said. "It's no wonder he went crazy. Especially with those blue walls!"

In the summer, on the way home from a fishing excursion with Andy, Jim stopped at the pet store in Drumheller and bought a canary to live in a cage he had found in the garage. What he didn't know was that the female canary does not sing, so Chirpy, not living up to her name, only sat quietly in her new home, suspended from the kitchen ceiling.

Later that summer, Ida struggled with a persistent throat infection. When she finally agreed to see Dr. Gourlay in late September, he suggested that she would only get better if her tonsils were removed. She arranged for a young woman, Jenny Steele, to cook, clean and care for the children while Jim was at work. The family seemed to cope well during Ida's week-long hospital stay, but that morning after she returned when she entered the kitchen, she looked around in confusion.

"Where's the canary?" she asked.

"The canary?" Jim queried now. "Oh God, I forgot about it."

He reached up and unhooked the cage, bringing it down to the table.

"It's just lying there with its feet up in the air!" Ida exclaimed. "What happened?"

"We forgot to feed it. It couldn't sing to remind us . . . I'll get you another one."

"Well, make sure the next one can sing for its supper!"

The next week, Mary Ann developed a painful swelling on the back of her neck and let her discomfort be known in no uncertain terms.

"It's getting redder and more swollen than it was yesterday," Ida told Jim. "You'd better run over to Andy's and get him to drive us to the doctor."

Dr. Gourlay examined the baby carefully, and then looked up to Ida.

"It was probably caused by a mosquito bite," Dr. Gourlay announced. "I'd like to put her in the hospital for a few days so we can treat the infection."

"I'll take you up to see her every day, Ida," Andy offered on the way home.

Two days later, as Jim was operating the electric coal-cutting machine, he hit the switch to stop it as his partner, Wilf Jones, had moved in front of the machine to clear some of the rubble created by the cutting blade.

"Bloody hell!" he hollered.

The switch failed to respond. The machine, propelled by hydraulic 'legs' kept moving forward as the sharp picks on the chain ate through the bottom of the coal face. Jim, knowing that Wilf was closer to the circuit breaker, ran to the front of the machine and jammed himself between it and the coalface.

"Wilf," he yelled, "get the hell out and pull the plug!"

Wilf responded immediately and the machine stopped, but not before Jim's shoulders were firmly and painfully wedged between the wall and the machine. Other miners rushed in and pried the machine back to release him. When they reached the surface, the pit boss told Jim that he had saved his partner's life.

Jim had only one response.

"Just give me something for the pain and I'll be fine."

"You'll get all of that at the hospital," the pit boss said.

They put Jim into the manager's car for the ride to Drumheller. Andy arrived at Ida's door in the late morning.

"Jim's been in an accident and he's in the hospital — but he's okay."

"What happened?" Ida demanded.

Andy explained briefly.

"The manager told me to come and get you," he added. "We can leave Cathy with Meg and if we're not back in time, she will keep an eye out for Jimmy after school."

"How are you doing?" Ida asked when she arrived at Jim's room.

"Healthy but dry."

"What did the doctor say?"

"Och, it was nothing, lass. My shoulder is badly bruised but there are no breaks. The pain is a bugger. How's my bonnie wee lassie?"

"I haven't seen her yet. We came straight to your room. I'll peek in on her and come back. Looks like you need to rest."

What could not be determined at that time was some kinking in Jim's carotid artery, which would eventually plug with cholesterol and require surgery when Jim was eighty-three.

Two days later, as Ida was taking a batch of bread out of the oven, she was thinking about the bad start they were having in their new home. First her tonsils, then Mary Ann's infection, and now Jim's accident. *Two in the hospital at once.* She looked down at five-year-old Cathy.

"Don't touch the hot pans," she warned. "I don't need you getting burned. I wonder what's happened to your brother. He's late."

Jimmy had started grade two that September and had to walk a mile along the tracks to school in the next village, Newcastle. As Ida began to prepare supper, her anxiety increased. *Where is that boy? There haven't been any trains this afternoon . . .*

"Get your coat on, Cathy. We're going to look for Jimmy."

They walked rapidly down the road to the Craigs' where Andy told Ida to wait with Meg while he looked for the boy. Andy soon returned with Jimmy, who was doubled over in pain.

"I found him sitting on the railroad track," he explained.

They called Dr. Gourlay, who arrived twenty minutes later. He carefully examined the boy.

"It's appendicitis, Ida," he announced. "I can take him to the hospital and operate tonight."

"I can't have three of my family in the hospital at the same time!" Ida said.

"Well . . .," the doctor thought for a moment. "These things sometimes ease up by themselves. Take him home and put him to bed. Call me if he gets any worse. Otherwise, phone me at eight in the morning."

Andy agreed to come by in the morning, see how the boy was and then phone the doctor. But when Jimmy awoke in the morning, the pain was gone.

"I'm feeling fine," he said. "Can I go to school?"

"No! You are going to stay home and take it easy. I don't want you to get sick."

Andy was pleased to see how well Jimmy looked.

"I'll call Gourlay and tell it's eased up."

Mary Ann was discharged three days later and Jim the following week. He was sore for a while, but since nothing was broken, he endured the pain and returned to work as quickly as possible. Jimmy had no further stomach pains.

Jim enrolled in night school again for the fall and winter term, taking courses that would complete the requirements for his fire boss papers. He and Andy studied mine safety, especially around the use of explosives and the testing for gasses such as the lethal methane often present in Alberta mines. Included in the training were supervision skills, as the fire boss supervised an area that often had as many as seventy men, the maximum he was allowed, working in it. The final testing took place over a day. The candidates had to answer a number of written questions, then sit before a panel for the oral part of the exam and finally participate in hands-on simulations. Both Jim and Andy easily passed and were qualified to work as fire bosses.

The economy was improving with the demands a war makes, and life was easier than it had ever been for Jim, especially having achieved his fire boss ticket. The mines continued full time during the spring and summer of 1941, and he decided he could now afford a car. He set out with Andy for Drumheller with $50 in his pocket and returned a few hours later, proudly and loudly honking the horn of his first car, a 1927 Willys. Ida and the children rushed out excitedly in response to the beckoning *oo-gah*.

"Come on, Ida, get the bairns ready and we'll go for ride!" Jim offered gleefully.

The car gave them a new freedom for shopping and outings that had up to now been limited by the generosity of others.

It was about this time that Jock Shearer, a former pit boss at the Commander mine who had moved to Hillcrest a year earlier, wrote to invite Jim and Andy to come and work with him. He said that the mines in the Crowsnest Pass were looking for skilled workers, especially fire bosses. Andy read the letter aloud as Jim drove back from the post office.

"What do you think, Andy?" Jim asked. "The mountains are a lot healthier place to be."

Dust storms regularly sandblasted the Drumheller Valley and violent thunderstorms still produced flash floods.

"Aye, and that's where the fishing is."

"Whatever will keep you happy," Ida replied when Jim broached the subject that evening.

They agreed that the men would drive down in Andy's car and check it out. If it was promising, they would get established and send for their families. The men settled on the Hillcrest-Mohawk mine in Bellevue and decided that it would be better to live in the little village of Hillcrest, nestled against a mountain across the Crowsnest River.

Once accommodation had been secured, Ida set about selling the house, which she did for $80, a profit of $20. This paid for a truck to move the two households of furniture with enough left over to hire a young man to drive her, Meg and the children to Hillcrest in the Elliots' Willys.

The first year in Hillcrest was refreshing for the whole family. Jim was quite content with the challenges and the pay that came with his new job. Ida and Jim made new friends quickly with Matt and Mabel Douglas and Meg and Alex McDade, and also reconnected with the Shearers. Cathy started school and Jimmy entered grade three.

Jim and Ida listened each night to the CBC News, which always began with, "This is Matthew Halton of the CBC reporting from . . ."

"That's Jean Moser's uncle," Jimmy announced.

The Haltons and Mosers ran the local general store.

Ida was busy knitting socks and mittens for the family, ceaselessly baking and managing all the little tasks that came up, such as helping to extricate a tongue that had been attracted to the white 'frosting' on the door hinges as fall gave way to winter.

During the summer of 1942, many weekends saw two or three carloads of friends heading off to Lundbreck Falls for a picnic. Jim and Andy soon taught Matt the fine art of fly fishing, and off the men would go, while the women and children visited and played together, all anticipating a huge trout feast to end the day. Fishing and friends helped keep life positive during the summer.

However, as fall approached, Jim was dismayed as he listened to Halton report on the bombings in London, the invasion of Russia, the entry of Japan into the fray, and the struggles in North Africa. Then, when Doll wrote to say that German bombers had got as far as Glasgow, his desire to join the army was renewed in earnest. Ida would not discuss it. She had made up her mind and she had already made her decision known.

Jim began to do things he had never done in their ten years of marriage. Instead of coming right home from work, he would stop off at the miners' club for a beer and a game of darts, hoping to get Ida angry enough to relent and finally tell him to get out and join the army. Still, he did not like coming home late and eating supper alone, and he soon realized that this was not working — Ida refused to confront him. One evening after a lukewarm supper, he decided to be direct.

"You're not going to fight with me, are you?"

"No, I'm not," she replied tersely. "You know where I stand."

"But Ida, the kids are getting older now, and Jimmy's a big help to you. I really want to do my part to get rid of Hitler and make the world a better place for you and the children. If someone doesn't stop him, he will keep going 'til he gets here."

He became relentless in his cajoling and, because Ida remained resolute, their life together became unpleasant. When Stella visited in October, she immediately sensed the tension and asked what was wrong. Ida and Jim each explained their position.

"I love you both," Stella told them. "I don't think you two will make it unless you can agree on this. Ida, sometimes you have to let go. The war won't last forever."

"I don't want him killed, Stella. It's hard enough having Pete over there."

"I could just as easily get killed in the mine," Jim offered. "I don't want to be branded like those zombies in Quebec."

The three of them sat in silence.

"I don't know what to do," Ida said finally. "Can you wait until the new year?"

"If you want me to, I will," Jim said.

He reached out and pulled her close.

Though the next two months flew by for Ida, they dragged slowly for Jim. Now there was different tension in the air as they savoured their days together. The army was never mentioned, and the energy of the children at Christmas carried them through a season of unspoken thoughts.

Jim left on January 2, 1943, for Calgary where he joined the Royal Canadian Engineers and was assigned the regimental number, M105894. He was immediately shipped to Camrose for basic training.

Chapter Nineteen
You're in the Army Now

"What do you mean you're sending me home?"

Jim didn't want to believe what he was hearing.

"We've had a letter that says you are needed in the mines and therefore are exempt from military service," the officer replied.

"Who the hell would write a letter like that?"

"I don't know. But I have my orders."

"Bugger that, I won't go!"

"As far as the army is concerned, you are discharged as of June 30, soldier. Dismissed."

Jim snapped a sharp salute, did a quick about-face and marched directly to his barracks. He was alone as he removed his uniform, changed into his civvies and began to pack his duffel bag, which was fortunate, as his ill-humour would not have tolerated any idle questions. He picked up the army-green scrapbook with the gold-embossed maple leaf on the cover, read the words, 'Snaps and Scraps — My Life in the Army' — a gift from Ida. Well, that was short. He opened it to the first page.

Arrived in Calgary, 4 Jan 1943. Stationed at Mewata till Jan 29.
Arrived Camrose 29th Jan to start Basic Training. Passed First Aid (A).
Feb 14th promoted to Acting Lance Corporal.
9th Feb Received my first break, had 60 hours leave, had a swell time at home.
Finished basic training March 27.
Left Camrose March 30.
Arrived Chilliwack April 1, had to walk six miles to camp in the rain.
Started advance training April 5. Weight 152 lbs. Gained 4 lbs.

Ida was overjoyed when Jim appeared at the door on July 2, 1943. Then she saw the familiar jacket and pants he was wearing.

"What's happened?"

"I'm discharged. Some bugger from the mine wrote to say I was needed here."

"But that's swell news, Jim!"

She threw her arms around her civilian husband.

"Well, I'm not happy about it," he said, stepping back. "Someone will have to answer for this. I want to be in the army."

"I'm glad you're home."

The next morning Jim arose early, had his bowl of porridge and strode off to the mine.

"What the hell is going on here?" he demanded of the mine manager.

"I need good men and I thought you would be happy to be out of the army," the manager replied, smiling at Jim.

"*You* wrote that bloody letter? I joined up because I wanted to get out of the mines, you silly bugger. You didn't even ask me."

"Well, you're here now, and I need a good fire boss. You may as well start work tomorrow."

"The hell you say," Jim quickly replied. "I'm not going to work."

"You have no choice. The government has decided that miners are essential to the war effort."

So Jim endured the mines and Ida endured Jim's ill temper for the next eleven weeks. He complained at home and was belligerent at work until the manager finally relented and sent off the necessary documents to the military authorities. Jim celebrated Jimmy's tenth birthday on the 19th, and was back in Chilliwack on September 21st.

With Jim gone, Ida kept busy cooking and washing for the three children and received a great deal of emotional support from the Craigs and Douglases. Neither Andy nor Matt wanted to enlist and both were content to do their bit in the mine. Jim wrote to Ida often and always included postcards for the children. The army provided the cards, which featured a painting entitled *Canadian Soldier* by Lilias Torrance Newton, a prominent Canadian artist. Jim's words to Jimmy were always the same, "Be a good boy, look after your mother. Stick in at school."

One day after the children had returned to school, just following lunch, and with Mary Ann settled in for a nap, Tony Lessen, the landlord, knocked at the door. Ida was loath to open it as she did not like being around this man. Twice in the recent past when she had gone to use the shared outhouse, she had found him sitting in it with the door open. As the door faced his house at the bottom of the lot, it was not possible for her to see him inside as she approached.

"Missus, I have some pictures of my family in the old country," he said. "Would you like to see them?"

Against her better judgment, she opened the door and led him down the long back porch and into the kitchen. Displaying the pictures, he began to crowd in on her and she slowly edged backward across the room. *My God, he's not interested in pictures.* She began to panic. If she kept going, she would end up in the bedroom — the last place she wanted to be. *God, help me!*

Suddenly the screen door burst open.

"Mommy! I forgot my orange," Jimmy announced as he ran into the kitchen.

Tony stepped back and Ida quickly grabbed her son and pulled him close. Her heart raced as her mind sought her next move.

"Get out of this house!" she screamed.

Tony looked at her, then at her son and quickly turned and fled out the back door. As she slowly released the bewildered boy, Ida began to shake.

"God sent you home, Jimmy. I don't want you to go back to school yet."

"What's the matter, mommy?"

"I don't want that man around."

She slumped into a chair and sat quietly for a few moments, breathing heavily. The boy had never seen his mother this upset before and just stood watching her. Then, slowly rising, she smoothed down her apron, reached for an orange and handed it to him.

"Come straight home from school and make sure Cathy is with you. Hurry now or you'll be late."

When Jimmy was gone, she stuck a knife in the doorjamb and made a cup of tea. The cup rattled in the saucer as she sipped and reflected on what she needed to do.

After school, she took the children over to the Douglas home and told them what had happened.

"Matt, I need to move," she concluded. "Now!"

"You stay here with Mabel while she gets supper ready and I'll see what I can do."

With that, he got into his car and began driving up and down the narrow streets of Hillcrest looking for a 'For Rent' sign. He was fuming as he thought of Tony and how lucky that scum was that Jim was not around. Finally, he found a two-bedroom house for rent close to the centre of the community and rushed home to tell Ida. She and the children spent three more days with knife-in-the-door security before Matt, Alex McDade and Jock Shearer could get the family's furniture moved to the new place.

In spite of the fact that Ida had a number of good friends for support, she was extremely distressed by the incident and was not sure she wanted to spend the rest of the war in Hillcrest. She needed Jim. She missed her family.

One morning in September, Matthew Halton reported that the Canadian troops had landed unopposed in Italy the previous day.

"We have crashed the gates of Europe," he said.

As Ida listened, she thought of Pete, knowing that he would probably be in that invasion force, since he had been involved in the push through Sicily. She wondered how her mother was coping and realized that she really would like to be closer to her family. She wrote to her parents, who were running a rooming house at 717 Hamilton Street in downtown Vancouver, and told them she wanted to move away from the Crowsnest Pass.

"It would be easier for Jim to visit us if we lived out there."

Her father replied and suggested she could come to the coast and live on the acreage he owned in Surrey.

"Mary and Carl are living there now," he explained, "but you can move in as soon as they are settled in their own place. Come as soon as you can and stay with us."

Once more, Ida sold the furniture and the car — she still did not know how to drive — and packed their remaining belongings. The children had mixed emotions as they left to go to the big city. They not only left friends behind, but there was no room for a dog at Nonna and Nonno's and they had to leave Rover behind. Dickie Thomas was excited about being the one to get the dog, and promised to take good care of him.

Ida and the children arrived in Vancouver in mid-October and spent three weeks with her parents in the rooming house Frank had leased, in turn renting out rooms to single men. Angelina did the laundry for their tenants but meals were not included. Because the apartment was not large, the family spent a great deal of time was outdoors exploring the neighbourhood and Stanley Park with their *nonno*. The city was bigger than anything they had ever seen and they were afraid to go too far from home. Jimmy and Cathy registered in the local elementary school and timidly walked back and forth together each day.

When Jim got a leave at the end of October, Ida and the children took the train to Mission to meet him and they all travelled back to Vancouver together. It was exciting for the family to see him smiling and in uniform. Mary Ann was able to get some of the special fatherly attention that had been hers prior to her daddy joining the army. The other children were happy to be seen walking in public with their soldier father. They always stood just a little straighter and smiled broadly as they approached the street photographer. Jim would take the ticket from the man and, without missing a step, keep marching along Granville Street with his proud children.

"Can we get this one, Daddy?" Cathy would ask each time.

"Aye, I want to have some good pictures of you all."

The leave was over too quickly, and with Jim gone, Ida decided that she and the children could not continue living in the crowded rooming house. Mary suggested that they move in with her, Carl and their three children on the acreage.

"It will only be for a month until we finish the house down the road," Mary urged.

Ida agreed and moved the next week. Jimmy and Cathy enrolled in the one-room Green Timbers School in Newton.

In mid-November, Jim got a two-week furlough. Once more, he and the family were living with relatives. They had little privacy in the cramped farmhouse, but they appreciated the time together, and Jim was determined not to raise a fuss with Mary about anything. He was successful until the morning she decided to make cabbage rolls.

An early riser, she had a full roasting pan of her delicious creation in the oven by 7 a.m. Unfortunately, the zesty odour emanating from the stove hit Jim's volatile gag reflex; he awoke with a start, dashed into the yard and immediately threw up. Never one to mince words, he returned to the kitchen and immediately confronted her.

"There's something really rotten around here. What the hell are you cooking?"

"Cabbage rolls for supper tonight. I decided to make a treat for everyone."

"That's no damn treat! You should toss out that bloody foreign cooking."

"I should toss you out, you stupid Scotchman!" Mary replied angrily.

Ida burst into the room.

"Jim, get some fresh air," she ordered.

With that, she handed him a shirt and escorted him outdoors, where he could cool off and clear out his lungs with a cigarette.

Ida did not like living in the Green Timbers area after Carl and Mary moved to their own place on a lot farther down the road. Although she had spent so much of her early life in remote houses where she couldn't see the neighbours, this was not what she had in mind when she left Hillcrest. Now her social life centred on Mary and her family, and the children complained about the long trek to the one-room school. The bus service was poor and walking became the only real alternative, but New Westminster was over four miles away across the Pattullo Bridge. Ida had to rely on Carl to drive her to shop for groceries. Loneliness overwhelmed her at times and, as she sat listening to the radio with its constant barrage of war news, she longed to talk to someone who would understand the feelings that were coming over her again. Mary told her to buck up, stop whining and keep herself busy cleaning the house.

On December 4, 1943, Jim was sent to Hamilton, Ontario, on a six-week course in engine maintenance. That Christmas was the first he and Ida had been apart in their eleven years of marriage. He kept busy — fully involved in the intensive program — but he missed his family, and on Christmas Day, as he ate what the Army called a turkey dinner, he wondered about his decision to enlist and if this was worth it.

The gray winter rain added despair to Ida's feelings of isolation, which she shared in a long letter to Stella.

I have just had my worst Christmas ever. This place is so depressing. All it does is rain and I don't see other people. There is nowhere to go.

Stella replied a few days later to tell her that Fabbi had found a small cottage and moved it onto their property directly across the road from their own home on Okanagan Lake. He was "fixing it up for Ida to live in." It would be ready by the end of February. A ray of hope broke through Ida's clouds of gloom. She accepted the offer without hesitation and immediately wrote to tell Jim her good news, her letter arriving not long after Stella's.

He finished his course on February 4th, and passed with a mark of 88 percent. Upon his return to Chilliwack, he was granted a forty-eight-hour leave and rushed home to his family to enjoy Ida's home cooking.

After a hearty meal and the children were off to bed, Jim and Ida sat down for a cup of tea.

"It's so good to be here with none of your family around," Jim said.

"Things will be better," Ida sighed. "We won't be here much longer."

Chilliwack was close enough to Newton for Jim to get home for the next two weekends. Then he took ten-day compassionate leave to help the family move to Mission in the Okanagan. Jim and Carl drove a borrowed truck that contained the boxes of their belongings, Ida and the children travelled by train to Vernon where Fabbi met them.

"It's perfect," Ida exclaimed.

She saw the cabin sitting in a tree-lined meadow on the bank of Mission Creek, across the road from Fabbi's house. She hugged Simon.

"Oh, Fabbi, you're wonderful."

"We want you to be happy, Ida."

"What are you building, Uncle Simon?" Jimmy asked, pointing at the two-by-four frame extending from one end of their new home.

"There are only two rooms here, so I am going to add two more as soon as I can get at it," Fabbi replied.

"Don't hold your breath," Stella whispered to Ida. "He has so many projects on the go. I hope he will finish this faster than some of the others."

Sure enough, Simon was not able to work the extension into his schedule during the duration of Ida's stay there.

Jim and Carl arrived late in the afternoon and unloaded the truck before sitting down to a dinner of fried chicken at Stella's.

With Jimmy and Cathy once more enrolled in a new school — their fourth that term — Ida and Jim busied themselves with getting things in order, while Mary Ann explored the meadow and chased the two goats that Simon had bought for their milk supply. In preparation for Jim's extended leave, Ida had saved meat and butter coupons. "You know how Daddy likes to see his teeth marks in the butter," she had reminded the children during their period of easing up on the rationed food.

Jim was happier than he had been in months and Ida wondered if life could get much better than this.

Then the telegram arrived: "Leave extended two more days prior to embarkation."

"What does that mean?" Ida asked.

"I'm being shipped overseas."

"Oh God, no!"

"Ida, hen, we knew it was coming."

"Promise me you won't do anything foolish and will come back safely."

"I'll be back, sweetheart."

Jim left Chilliwack on March 21 for the staging area in Camp Debert, Nova Scotia. On April 29, following another month of intense combat training, he and his mates boarded the troop ship that would take them to war. He was sick for much of the trip — nothing like the luxury he had experienced on his way to Canada — but his spirits brightened when the ship slowed as it manoeuvred around some islands.

There was something familiar about the scene. He had seen all of this before when he had watched from the deck of the Montrose as his homeland receded. *We're heading for Greenock.* He was home in Scotland! Memories of his family washed over him. *Will I get to see my mother?* However, immediately upon docking, the troops were transported to the station where they boarded the train that took them to their base in southern England. *I'll get a leave*, he decided.

His first blue airmail letter to Canada contained the news that he was stationed somewhere in England, but Ida could not read the place as the censor's black pen had carefully expunged it. They did not erase the fact that he was living in a tent. He assured her that everything was going well but he really missed her, the children and her home cooking.

Ida decided to send a parcel to Jim with some cookies and warm socks. Jimmy asked if he could help and soon had made his first batch of muffins. She carefully wrapped a few and included them with the other items, notifying Jim in the enclosed letter of the significance of the muffins.

Jim and the other troops gathered for the most-welcome routine of mail call. He was only too happy to hear his name called.

"M105894, Sapper Elliot, J," the mail dispatcher shouted.

Jim eagerly retrieved the small parcel and two letters and stuffed them into his chest pack as he headed for the railway station and off to a new camp. When he was finally seated on the train, he quickly opened his parcel, rummaged through it and found shortbread, gingersnaps and some very hard muffins that he decided he would not even try. The train was just clear of the station when he threw the homemade bricks out of the window, popped a piece of shortbread into his mouth and eagerly opened the letter. He turned quickly to the window and looked helplessly and hopelessly at the departing countryside, quietly savouring the shortbread and waiting for the tears to stop. When the train pulled into a station somewhere in the south of England, he mailed a letter explaining to Ida what had happened and asking her to tell Jimmy that they were great muffins.

Ida spent a great deal of time across the road and enjoyed her heated discussions with Simon, as she stood up to him in a way that Stella never would. That spring, she and Stella decided to plant potatoes and set out a string to mark the first row. However, the string was too loose and a slight breeze wowed the line, and each succeeding row followed the same curve. Later, when the first green shoots appeared, Simon, upset by the curvy potato patch, spent a whole day digging up the tiny sprouts and replanting them properly in *straight* rows.

On Tuesday, June 6, 1944, Ida dashed over to Stella's after the children had ridden off to school.

"Are you listening to the radio?" she asked breathlessly.

"Yes, come in."

They sat looking at the little green eye on the Marconi as the voice of Matthew Halton reported that the invasion of Normandy had taken place earlier that day. He had gone over with the first wave of troops, landed with them and then returned to England to record his message for broadcast to Canada. With a seven-hour time difference, the message arrived on the same day. He reported that the engineers were detonating land mines to clear the way for the combat troops. Ida, Simon and Stella looked at each other soundlessly, each wondering if Jim was in the midst of it all.

Jim Elliot was not in that first wave. Instead, the landing craft carrying his unit, the 5th Field Company of the Royal Canadian Engineers, was still slowly negotiating the churning English Channel. Jim's stomach protested the constant bobbing of the craft.

"Are you not going to drink your tot, laddie?" he said to the pale young man beside him.

"No, it's too strong," the young man whimpered, clinging to the side of the boat.

"I'll look after it for you. My stomach needs some settling," Jim replied.

He took the man's cup and poured the black rum into his own. He smacked his lips.

"That's not bad."

Some of the other young soldiers around him poured their unfinished drinks into his cup and, before long, he was no longer aware of his nausea. The noise of the guns from the large ships, the constant roar of plane engines overhead, the thunderous explosions on the beach and the steady drumming of machine guns announced to this second wave that the invasion was well underway.

It was the job of Jim's unit to remove underwater obstacles at the beachhead with bulldozers, move on with mine detectors to clear a wider path through the minefields and then construct bridges for the support vehicles coming up behind them. The first wave of infantry had been able to push the Germans back before they could blow up all of the bridges over the River Orne and the Caen Canal, but, unfortunately, these bridges were small and could not support the army's large transport vehicles, and so it was necessary to build Bailey bridges across these waterways.

The Bailey bridge, a portable pre-fabricated truss bridge, designed for use by military engineers, could span up to 200-foot gaps between supports. The components were small enough to be carried in trucks and no special tools or heavy equipment were required for construction. Once completed, the bridge was strong enough to bear the weight of tanks.

About ten days after the invasion, a wire photo on the front page of the *Vancouver Province* showed a group of soldiers in a landing craft as it approached the beach on D-day.

"My God, that's Jim!" Ida exclaimed.

She examined the profile of a grinning soldier in the bow of the boat.

"Look, Stella, that's Jim!" she repeated as she arrived breathlessly on Stella's doorstep, waving the newspaper. "What's he laughing about? The others all look so serious!"

A week later, a letter from Jim arrived and it confirmed that he had, indeed, landed in France on D-day and that they had run out of vomit bags in the landing craft so he had eased his nausea with "a wee dram."

Ida's letters to Jim always contained the admonition to "take care of yourself and come back in one piece." She usually asked the children to add a few words to the bottom of the blue paper.

Jimmy, caught up in the excitement of the war, wrote, "I hope you win a medal."

"You erase that right now," Ida ordered. "The silly fool might just try something stupid to get you one."

Letters during wartime needed to be censored in both directions.

In September 1944, the Canadian army experienced little resistance as it moved through Belgium and into Holland, where the Dutch anticipated their imminent liberation. The Germans were retreating north and east toward their homeland, and it began to look as though the war might be over by Christmas. Unfortunately, Allied supplies just could not keep up with the rapid advance, and their tanks were forced to stop due to fuel shortages. All supplies were still coming through Normandy at this time because the Dutch and Belgian ports, which had been virtually destroyed, were not yet free of German holdouts.

This situation contributed to a major military miscalculation in Operation Market Garden when, following an argument between General Eisenhower and General Montgomery, the latter went ahead with a plan to send thousands of British paratroopers behind enemy lines across the Rhine at Arnhem. The Germans counterattacked and soon the paratroopers were trapped, their resources depleted. The 5th Field Company joined the 20th and 23rd Field Companies to help rescue the desperate men.

On September 26th, under cover of darkness, Jim's company brought up heavy, twenty-foot, plywood storm boats, man-handled them up and over the wet and slippery dikes along the Rhine riverbank and into the water. German shellfire persisted, but when they started up the noisy Evinrude outboard motors, the sound gave the Germans a direction in which to focus their fire. As soon as the first three Canadian storm boats were launched, one of them received a direct hit by a mortar bomb, killing all three engineers in the first boat and another two in the second. The dramatic incident, illuminated by a burning factory behind the Canadian line, was witnessed by troops on both sides of the river.

Jim's crew got their boat into the water and headed across the river.

"This is hell!" he hollered to one of the Brits as they returned with a load of men from the far side of the river.

"Worse," the man groaned as he ducked low in the boat, mindful of the tracer bullets that indicated where the spray of machinegun fire from both sides of the river was heading.

Many of the British troops had been wounded and their pain-wracked voices added to the anguish of the rough crossing. As soon as their boats were unloaded, the Canadians headed out into the inferno again.

All night, it was back-and-forth under the non-stop barrage until about 2,100 of the 10,000 troops who had begun the operation had been rescued. But the light of the new day precluded any more runs, and those remaining on the far side of the river were either captured or killed in the following days.

In October, the Canadians were assigned the task of clearing the last German army holdouts from the sea coast of Belgium in order to free the ports. The work was non-stop, as the retreating enemy had destroyed all the bridges, and the engineers had to erect Baileys as fast as they could so that supplies could keep up with the advancing army. A few die-hard snipers hindered the advance, and when one of them killed the 5th Field Company cook, Jim volunteered to take over until a replacement arrived. It was a break from the wearisome bridge work. One of the men ladling food heard someone grumbling.

"Did you call the cook a bastard?"

"Do you call that bastard a cook?" the soldier replied.

Jim overheard them.

"Laddie," he snapped, "is it not to your liking?"

"I've had better."

"So have I. And we'll have it again after this bloody war is over." And dredging up the oft-heard words of his mother, he added, "Now eat what's in front of you."

Rest days away from the front were infrequent, but they provided an opportunity for leisure activities such as playing soccer. It had been some time since Jim had the opportunity to enjoy the skills of his youth, and in his eagerness to score his first goal in many years, he went for a ball near the opposition net and did not see the charging defenceman rush in to challenge him. The loud crack of boot hitting bone removed any doubt about the outcome — a broken leg.

"Bloody hell!" Jim exclaimed through the pain. "What did you do that for?"

The large fullback said nothing. Jim looked down at his leg and thought that this was probably the end of the war for him.

He was flown to the Cambridge Military Hospital in Aldershot, England, and remained there for the next six weeks while his leg healed. During that time, his brother Bill and sister-in-law Meg were able to arrange a trip from Tillicoultry to visit him and bring news from home. Bill loved the stories Jim told of life in Canada and said how much he would like to emigrate but did not have the money.

"Even with the war going on, we're not making much in the mines," Bill explained.

"Aye, and we've three bairns tae feed," added Meg in her broad Fife accent.

"Keep putting a little aside when you can," Jim offered. "It will soon add up. And you never know what it will be like when this bloody war is over."

"Could you give us a hand if we need it?" Bill asked.

"We'll do what we can when I'm back at work."

When Ida received the telegram of Jim's accident, she was overjoyed.

"Your father broke his leg in Belgium and is out of the war," she told the children excitedly when they returned from school.

"Did he get wounded by the Germans?" Jimmy asked.

"No, he broke it playing football. Isn't that great?"

Jimmy was not so sure. It wouldn't make a very impressive story for his friends.

After his release from hospital, Jim was given a leave for a trip to Scotland prior to rejoining his unit. It was sixteen years since he had left for Canada and he had so much to tell his family.

"Let me look at you," Martha said as she pulled him close. "You're awfully pale."

"I'm fine, Mother."

Jim had lost some weight during his convalescence and had not been able to get out for as much fresh air as he had grown used to.

"You're not the young lad that left here," his father said.

"A wife and three children and a war will do that to you."

"Sit down and tell us all about Canada."

"How are my brothers and sisters and their bairns?"

"They're all well," Martha said. "If they want to see you, they can come here. I'm no' letting you out of my sight."

Jim shared stories of his adventures in Canada, carefully avoiding some of the more painful parts that would surely disturb his mother.

"Life is good," he enthused, "and when I get back from this war, I'll be able to finally get out of the mines."

"What will you do?" his father queried.

"Farm."

Jim rejoined his unit in Holland early in 1945. As the weather warmed, the war started to cool down. The German soldiers, recognizing that Hitler's world-conquering dream was about to end, gave little resistance to the invading army. Jim's unit reached the village of Nordhorn on the German side of the Rhine River in time to hear the news that Germany had surrendered.

Following a few days of celebrations and clean-up, their unit moved back across the river to Hengelo in Holland where they were billeted with local people. The Dutch were extremely grateful to the Canadians for their liberation from the Germans and opened their doors and their hearts to welcome the soldiers into their homes as family.

Jim was billeted with the Posthuma family who had a little girl, Ellie, about the same age as Mary Ann. When he saw the poverty that had been their lot for so much of the war, he began bringing daily treats from the mess tent — white bread, chocolate, cheese and anything else he could lay his hands on. Most of the other soldiers were doing the same, and the officers gave tacit permission as there was no other means of aid available at the time. Each day when Jim returned, Ellie would be waiting for him to see what he had brought. However, the closer he got to the family, the more he longed for home. He knew that he would have some time to wait for repatriation as soldiers were returned home in the order that they had been shipped overseas so it might take six months for his turn to come up. There must be a way to speed up the process.

As soon as Simon had heard the news of the war's end, he suggested they all pile into the car and head for Kelowna. Jimmy picked up a cowbell and began ringing it out the window as they drove the two miles into town. When they got to Pandosy Street, it was jammed with people. No one seemed to know quite what to do except everyone wanted to express their extreme joy that the conflict was over in Europe. People laughed, danced in the street, one man sat down in the middle of it all and opened a bottle of beer — even though public drinking was against the law.

"I have waited almost five years for this," he shouted.

Later, when Ida read an article in the paper outlining the order of troops returning to Canada, she became very discouraged.

"I don't want to wait another six months to see him," she said to Stella. "There must be some way to speed up the process,"

"If there is, Jim will figure it out," Stella said.

Chapter Twenty
A Farmer He Is . . . Not!

Jim pulled out the pamphlet, *After Victory in Europe*, and re-read the pertinent parts. Circulated ten days after VE Day, it contained information on how the repatriation program of 'first in, first out' would affect him. It involved a combination of points for years of military service and once again, he did the math in his head. *Two for each month in Canada . . . 32; three for each month overseas . . . 51; add 20% for the children . . . still only 99. About half of what Pete has.* Jim realized that this would not put him very high on the list, but there was another option available in the 'Specialized Workers' section.

Back home, the country was already gearing up for post-war production and since many industries were powered by coal, miners could get an exemption. *If it will get me home faster — I'll do it! I'll spend time in the bloody mines just to be with my family again.* Jim filled out the forms and, instead of a long lonely wait, he was on a ship within four months of the cessation of fighting in Europe.

It was Simon's idea for all of the family to meet Jim at the military base in Vernon instead of waiting in Kelowna for the bus that would transport troops south from the end of the rail line. The eight of them had piled into the car, no one grumbling about not having enough room.

Just one more hour, Jim thought when the conductor announced "Vernon!"

As the train slowed for the station, he looked at the crowd waiting for their loved ones. It had been the same all along the line. Every station platform saw eager families waiting for their veterans to disembark while impatient soldiers stood in the aisles as their destinations were called out, crouching to peer out of the windows, each vying to be the first off. Since everyone was detraining in Vernon, Jim was not in a rush — there was plenty of time to get the bus and continue on home. He peered out the window as the aisles filled with the Vernon troops who wanted off first. *Holy God, that looks like Fabbi! It* is *Fabbi. And there's Ida and Stella and the children.* He quickly raised the window and waved.

"Ida, Ida!" he hollered.

As one, the family found the source of the excitement and responded with cheers and waves. Jim grabbed his kit bag and jammed himself between two soldiers.

"Sorry, laddie, my family's here!"

"So's mine!"

It had been almost a year-and-a-half since Jim's embarkation leave, and the bedlam of the reunion was pure delight as each person tried to get a piece of him and he tried to be there for all of them. Again, no one noticed how crowded the car was during the noisy ride home.

Once the children had settled down in bed on Jim's first night home on that early September evening in 1945, he and Ida cuddled on the chesterfield as they sipped tea.

"They'll send me to jail if I don't go back to the mines," Jim said. "I have to work there for at least six months. That was the deal in order for me to come back early."

"I'm just glad that you are home safely. We can live anywhere."

"The army will pay me for another two weeks as I get settled, but I'm supposed to report to the manager at the Mohawk as soon as I can."

"Let's move quickly, then. The children have just started back to school and it will be easier on them if we don't wait too long."

A week later, Jim set off for Hillcrest. Andy Craig insisted he stay with them while he looked for a house. Jim knew Meg was a wonderful cook and fun to be around when she was sober, so he suggested to Andy that there not be any partying while he was with them. Betty was happy to have 'Uncle Jim' around as her mom didn't drink very much when he was there. They all relaxed while Meg cooked, playing her favourite Christmas music on her RCA Victor gramophone.

"I don't care if it is September. I like it."

The manager of the Mohawk mine was pleased to have this strong-willed, hard-working Scot back, so did not comment on Jim's behaviour prior to his departure two years earlier. Jim began work immediately as a fire boss. Once again, he would walk through the valley from Hillcrest to Bellevue each day, pick up his lamp and descend the stairs to the beginning of the slope, walking down it into the bowels of the mine. Jim liked the responsibility of being a fire boss — it was less laborious than digging or running a machine and the pay was better — but it was still the mine.

Within a week of his arrival, Jim rented a two-storey, three-bedroom house owned by the Taber family and sent for Ida and the children. It was the largest house they had ever lived in and, for the first time in his life, Jimmy, who had just turned twelve, not only had his own room, but it was upstairs — another first. The family had a sense of coming home when they returned to Hillcrest, as the children and adults alike quickly reconnected with old friends who were delighted to see them. Again, the children had mixed feelings about returning, as Dickie told them that Rover had died after being hit by a car.

Ida and Jim easily moved back into the social circle that included the Douglases, Craigs, McDades, Shearers, Bains and the Browns — all Scots. Once again, Ida buried her Italian roots under Scottish customs, songs and food.

After almost three years in the army, out in fresh air, Jim was not happy being underground again. However, he suppressed his negative feelings and accepted the fact that this was the price he had to pay to be back with his family. While he listened to the catch-up stories from four bubbling springs of refreshing memories, he quickly dismissed questions of what it had been like overseas.

"It was long and hard and I'm happy to be home. What else happened when I was away?"

The flow resumed.

Still, as the shortening days heralded the coming of winter, Ida was regularly wakened in the middle of the night by Jim mumbling 'Take cover!' or 'Is that ours or Jerry's?' This always coincided with the passing of the Trans-Canada Air Lines mail plane overhead on its way from Vancouver to Toronto. The drone of the engines triggered something deep inside of Jim that caused him to break into a cold sweat. Ida, frightened by this strange behaviour, would wake him.

"Are you okay? Is it about your war experiences?"

He had no memory of the dream and was not sure what was happening.

"I don't want to talk about that. I've told you that."

He would roll over and drift back into sleep. But it was not as easy for Ida to settle into sleep again and, as the weeks passed, she began to worry about Jim's health. His reaction to the plane eased off as time went on, but he still would not talk about the war except to tell the odd humorous anecdote or describe the countryside and the beauty of the farmland that had been badly torn up by warfare.

"I admired the strength of the farmers we met when we were moving across France and especially when we got into Belgium and Holland. They had nothing. The Germans took all the produce and animals for their own use, and then they flooded the land with sea water."

"How awful for the people."

"Yes, but as soon as they were liberated, they began to work the ground again and start planting." He paused, rubbing his chin between his thumb and forefinger and looking into Ida's eyes. "There is something about farming."

However, most of his war talk centered on Holland and the people there. Ida had written the Posthumas to thank them for looking after Jim and that had resulted in regular correspondence back and forth to Holland.

In January 1946, Jim took two days off work to travel to Calgary for his formal discharge from the army. As the officer in charge read Jim's documents, he reminded him that he was still under obligation to work another three months in the mine.

"I know that, laddie, but after that, I want to get my own farm."

"In that case, while you're here, it is important that you talk to the people in the DVA and VLA."

Jim was of course familiar with the Department of Veterans' Affairs and the Veterans Land Act, but unsure what the connection could be.

"What for?"

"They will let you know what credits you have because of your service and how the government can help you get started."

Jim returned home with the necessary forms and told Ida of the visit to the VLA office, and how they loaned money to veterans at a low interest rate if they were buying a house or property. He filled out the forms and mailed them two days later.

One afternoon in late February, when Jim returned from work, Ida had an announcement.

"There's an official looking letter for you."

"Let's have it."

He tore the envelope open and quickly scanned the letter.

"I'm going to be a farmer! Listen to this, 'Arrangements have been made for you to begin work as a trainee on the Ivan Meyer farm in Coaldale beginning April 1, 1946.' Ida, it's happening!"

"That's just eight miles from Lethbridge. Where will we live?"

"Um . . . let's see . . . here it is . . . they have a place to stay on the farm and DVA will supplement my wages beyond what Meyers will pay. After a year, he will assess my farming skills and then we get a place of our own."

At the end of May, the family was on the move again and quickly settled into their farm accommodation, a two-room cabin with a construction trailer beside it that would provide a bedroom for Jimmy and Cathy. For the first time in their lives, the children experienced the bouncing, raucous daily trip to and from school in a yellow bus but, as the others all seemed to know each other, the three newcomers sat quietly at the front.

The Meyers lived in a two-storey house across the road. Ivan and his wife Betty had a three-year-old boy named after his grandfather, Bill, who also lived with them. Old Bill, as they called him, had cultivated most of the ten sections of the family farm for crops, but had kept three sections along the riverbank as grazing land for their small herd of beef cattle.

Jim had never driven a tractor before and it took some time for him to learn how to manoeuvre it while towing something. His first lesson was a bit costly when Ivan asked him to drive a load of hay around to the barn. Jim cut the corner a bit tight and the wagon clipped off a piece of the barn before he could stop the tractor. Fortunately, Ivan, a laid-back easygoing man, just laughed it off, telling Jim he would have to develop some carpentry skills as well as those necessary for farming.

Jim learned the different drill settings for oats, barley and wheat and drove one of the two tractors to assist Ivan with the seeding. He enjoyed the outdoors, the fresh air and the warming spring sun on his body. This is what he had long dreamed of — being a farmer. Meanwhile, Ida planted a large vegetable garden in the plot Ivan had tilled for her behind the cabin.

A neighbour brought over a puppy, a German shepherd-collie cross, who was promptly named Rover, in memoriam of his namesake. His herding instinct emerged as he and Jimmy brought in the few milk cows each evening before supper. Ivan and Old Bill looked after the milking and did not press Jim to participate after he had explained his previous experiences with cows.

Life was good for the family, and as the summer heated up, Ida discovered she was pregnant. Ida preserved fruit and vegetables in wide-mouth mason jars. It was hot work as the process involved boiling each batch for three hours on the coal stove in the small kitchen. She rested outside in the shade while the preserves boiled, sipping water to ease the nausea that accompanied each pregnancy. She served fresh vegetables daily, preserved each variety in its season, and then — another first — she froze food.

Betty had suggested this.

"We keep all of our meat and most of the vegetables in a locker at the cold storage place in Coaldale. We go in each week and get as much as the freezer compartment of the fridge will take."

Ivan gave them a side of beef all cut and wrapped and Ida stored it in her locker. The Meyers were generous about picking up items for Ida in Coaldale and periodically would take her shopping there.

During the hot summer months, Jim spent his days on the tractor, pulling the harrows round and round the fields of summer fallow. He was exhausted and dry when he returned home each evening after a long day in the prairie heat.

"This is worse than coal dust," he said to Ida. "It coats the teeth and the lining of my mouth and throat."

"Maybe it's because you are smoking and not chewing snuff like you did in the mine. Are you sure you want to keep doing this?"

Ivan had told her that he thought Jim just didn't have what it takes to be a farmer.

"Of course I do. Ivan told me that a beer works well with the dust."

He went out to the cool root cellar, returned with a bottle of Lethbridge Pilsner and took a long drink of the cool, refreshing liquid.

"This is much better than snuff," he proclaimed.

Ida turned to get the supper from the stove to the table. She swallowed the urge to say something to Jim, but did not want to curb his enthusiasm. *He's been through so much to get here.*

When threshing time came, Jim's need for a beer a day did not diminish as the dust from the combine filled the air. His only reprieve came when he drove the loaded truck to the grain elevator, with a cigarette stuck to his lips and smoke drifting out of the open window. Ivan usually took that time to tinker with the combine or the tractor. This part of farming befuddled Jim, as he had little experience around anything mechanical except the coal-cutting machines, and there had always been an engineer in the mines to maintain them.

One day when Jim came in to supper, Ida was at the table with her notebook in front of her.

"Stew's in the oven. I canned some beets today, so the stove was hot anyway. I'll be finished in a minute," she said.

Jim opened a beer and sat down across from her.

"What are you doing?" he asked.

"Making a list of all of my preserves. Here, have a look."

Jim picked up the pages spread in front of her and began to read.

"Ida, there's over a hundred and fifty jars here and that doesn't include the stuff you've frozen. I didn't realize how much you've done."

"That's just vegetables and fruit, here's the list of my pickles and jams," she said proudly.

"You're a wonder, Ida," Jim said, reaching for her hand across the table.

"Well, with another baby coming, I don't feel like doing much more."

"When do you think it will come?"

"I think it will be early spring."

"That's around seeding time. I'll be busy."

"You know me, I'll manage."

Jim was relaxed on the farm and eased back into the old patterns that included his ability to offer special cures for some of the ailments that were common with children.

"My mother had all the tricks," proclaimed the would-be doctor.

For an upset stomach, he would coddle an egg in a mug set in a pot of boiling water, stirring it constantly as it thickened into bright yellow custard. It was a great comfort food and the most requested.

"Daddy, I'm not feeling well. Can I have an egg in a cup?"

These occasions were among the few times Jim did any cooking.

For colds, he would make a hot toddy to be drunk prior to bedtime. He stirred honey into a cup of hot water and then added a tablespoon of the magic ingredient — whiskey.

"That was good, Daddy. Can I have another one?"

"That will do you. Now go to sleep."

An earache called for a different approach. He would light a cigarette, fill his mouth with smoke and gently blow it into the offending ear. It usually only took two or three puffs for the nicotine to mask the pain. He knew he would have made a good doctor.

During the winter of 1946–47, Jim spent most of his time learning about farm machinery and its maintenance. Ivan stored each item of equipment in a large shed and then, one by one, moved them into his shop for servicing.

Ida did not like the isolation of the farm and, being six months pregnant, she did not want to be out in the country when the baby was born, but she continued to keep her feelings to herself. She made the best of it as the family celebrated Christmas and moved into the new year.

"My year is almost up," Jim said to Ida one Saturday morning in February. "I'm going to have to start looking for a farm to buy."

"That's a big step. Are you sure that's what you want to do?"

"I told you that when I came back. The VLA guy says that once I find a farm I like, they'll begin the financing process."

On his day off each week, Jim hitchhiked to locations advertised in the *Lethbridge Herald*. Some days, he walked as much as he rode and came home discouraged. The worst was the long lonely seventeen miles to a farm beyond Picture Butte that turned out to be too run-down to even consider. Finally, he found a place in the country west of Granum, twenty miles north of Fort McLeod. He returned late on a Sunday, exhausted yet triumphant.

"You'll love it, Ida. It's in the rolling hills on the way to the mountains. You have to see it."

Ida just listened as he filled her in on details about the distance to Granum and the arrangements for schooling.

The next weekend, Simon and Stella, who were visiting Romeo and Stan in Lethbridge, drove out to see the Elliots. When Jim told them of the farm, Simon insisted that they go and look at it. Betty agreed to watch the children while they were gone. Ida, in her seventh month of pregnancy, did not look forward to a trip over country roads but agreed to go without complaint.

When they arrived at the farm an hour and a half later, Ida looked at the few buildings huddled close together on the vast landscape with no others in sight.

"I don't feel very well," she said. "I'm not getting out."

Stella remained with her while the men scouted the place.

"Stella, I don't want to live here. There isn't even a school bus to this God-forsaken place. The children will have to stay in town during the week. And it's so far away from people."

She began to cry.

"Does Jim know?"

"No. And I don't want to make it harder for him. But, Stella, Ivan says he doesn't have a farming bone in his body."

She could not stop the flow of tears.

As Jim and Simon explored the site, the Italian entrepreneur encouraged Jim to take it, assuring him that even though there was a great deal of maintenance to do, he could handle it.

"You could turn this into a real money-maker."

Jim returned to the car and saw Ida crying.

"What's wrong, Ida?" he asked.

"I guess I don't feel so good."

"Jim," Stella interrupted, "Ida doesn't want to live on a farm. She thinks it will be too much for you and too lonely for her and the children."

"Is that right?" Jim asked Ida, a bit bewildered.

"Yes," she replied as she wiped her eyes.

"Well, why didn't you say so? Let's go home and forget it."

He walked around the car, took his place in the front seat, and leaned out the window.

"Come on, Fabbi. Get in!" he hollered.

It was a long, silent trip home.

Stella helped Ida prepare supper while Jim and Simon paced the yard, absently watching Jimmy and his dog bringing in the cows.

"I guess it's back in the mines for me," Jim said finally. *Will I ever get out of this rut?*

Over supper, Stella suggested that Jim could probably stay with Romeo and Beulah while he looked for a job, and agreed to make the arrangements. He had no difficulty getting on at Number 8 mine. The problem was housing. There was nothing available in Lethbridge in that post-war period so he ended up renting a little house in Coalhurst, the former mining town six miles west of Lethbridge on the main highway. As a number of the men living in the village worked at Number 8, he was able to arrange a ride to work.

He did not like Number 8, a wet mine. Its shaft sloped down and under the Oldman River, and there was constant seepage of water into the work areas. The pumps ran continuously in an attempt to keep conditions manageable. Jim had no respite from the water as he moved from room to dripping room stuffing the shot material into the prepared holes, carefully stringing out the wire to a safe spot around the corner from the face, hollering 'Shot firing!' and then detonating the blast that brought the coal down. Each day as he returned for yet another shift, he entered the washhouse, untied the rope to the pulley, slowly lowered the hook holding his work clothes, reluctantly shed the clean dry pants and shirt he had worn from home, hooked them onto the pulley system and raised them to the ceiling of the washhouse. He then squirmed into the still-damp clothing he had worn the day before. As he hung his clean clothes on the hook, he often thought about the many days he had spent in his wet uniform slogging across Europe, but in spite of the hell of war, at least that had been outdoors.

The children had become used to moving and had accepted the fact that they would start at a new school in the midst of a term, March this time.

Jimmy made friends with the McLaren boys next door and soon became part of the Saturday morning water gang. With no running water in their homes, people relied on the town cistern that was regularly filled by a truck from Lethbridge. One of the fathers had built a cart to hold a forty-five-gallon drum that sat between a pair of wagon wheels. There were handles at each end so that four young boys could manage it. They pushed the cart uphill to the cistern, took turns pulling down on the long handle of the pump until the drum was full, and then negotiated the heavy load down the hill toward their homes where they then bailed out the water. It took all morning to make the four trips to fill each family's storage drum.

Life changed dramatically for the family when William Ian Frank, other-wise known as Billy, was born five weeks prematurely on April 14, 1947. Jim mellowed a bit when the new baby arrived, and he helped Ida with some of the cooking. Cathy and Mary Ann tested their mothering skills by watching the infant when Ida washed clothes or did other household chores. Billy proved to be a bit of a problem for the shy Jimmy when some of the girls in his class at school heard about the birth and asked if they could walk home with him to see the baby. It did not help to have Ida tell him that she thought they were more interested in being with him than with his baby brother.

Other than the Roman Catholic Church, the only church in Coalhurst was Pentecostal. Ida and the children began attending when the weather warmed and Jim was comfortable looking after the baby for an hour or so. She soon discovered, however, it was nothing like the sedate United Church or the formal Catholic Church of her childhood. The services were very emotional and some of the regulars had ecstatic experiences that left them writhing on the floor and speaking in tongues. After three Sundays, she stopped going, although she continued to send the children to Sunday school. Mr. Robertson, the minister, was a Scot and loved to drop by and chat with Jim who was not bashful about letting him know why he did not attend the church.

"I had enough of that as a child to last me a lifetime. Anyway, church is for bad people and I'm no' bad."

The small house was not adequate for the growing family and when their neighbours, the Deaks, turned a vacant store on the main street into an apartment, the Elliot family moved in. It had almost as much room as the Hillcrest house but there was no heat in it except for the kitchen stove. This was not a problem until the winter when the temperature dropped well below zero outside and almost as low inside. Ida got up early each morning to get the fire going in the stove, drape the children's underwear over the oven door and, when the clothes were warm enough, take them to the bedroom so the children could dress under the bedclothes. This situation angered Jim, but his arguments with the landlord over getting the heat fixed did not result in any change. It was just one more thing to add to his frustrations.

With the first hints of spring in 1948, Jim felt the urge to move again. He had endured the wet mine for a year but needed to get out.

"Ida, what do you think about us getting acreage through the VLA?"

"What would you do with that?"

"We could raise chickens and sell them and their eggs."

"Ah, Jim . . ."

"There's good money in it."

"As long as it's not way off in the country somewhere."

"Most of the chicken farms are close to Lethbridge."

"If you think you can do it."

As soon as the weather got warmer, Jim began searching the paper. On weekends, he would hitchhike into Lethbridge and then walk through the city to the advertised locations. He found a place right on the edge of the city that already had a number of chickens with adequate coops and pens. His friend, George Armit, had a car and he drove Jim out to look at the place one more time before he contacted the VLA office. Jim filled out all of the forms and was assured by the man in charge that it looked like a real possibility and he would be notified in a few days.

When the letter arrived, Jim tore it open.

"Here it is Ida! They say that . . . What the hell?"

"What is it?"

"They've turned down my application."

"Why?"

"I don't know, but I'll bloody well find out."

George Armit was a World War I veteran who had worked in the DVA office, and agreed to go to the VLA office with Jim. In the meantime, George did some research. When he and Jim met in front of the government building, he told Jim that one of the men who worked for the VLA had purchased the place for himself.

"This is not right!" Jim hollered.

A Farmer He Is . . . Not   261

He turned and dashed into the building with George struggling to keep up. When he arrived at the office, he leapt over the desk and grabbed the man.

"You bloody crook! I ought to throw you out the window!"

George quickly grabbed Jim's arm as the clerk continued.

"Calm down, man. This won't get you anywhere. Now, what's the problem?"

George stepped forward as Jim stood there seething.

"This is Jim Elliot," George said. "He had an application for a small holding denied."

"Not denied! Stolen! Someone in this office was given the property."

"I'm sorry, Mr. Elliot. I don't know how it happened."

"Do you think I was born yesterday, you stupid bugger? You sit behind a desk and help out your friends while the rest of us were getting shot at."

"Jim, let's go for a coffee and see what we can do," George urged.

Jim, knowing that he was no match for a crooked system, ripped up the letter he had so carefully carried with him and threw it at the cringing official.

"It looks like I'm stuck in the mines," he told Ida when he returned. "There is no way for me to deal with the people in that office."

"I don't want to spend another winter in this freezing house," she replied. "Can we move to Lethbridge?"

"You know we can't. I look at the paper every day and there is nothing."

Jim despaired over their situation. Every road ahead of him seemed to end at a black coalface in an unlit mine.

"There has to be *something*."

As they sat silently, Jim closed his eyes, searching for pictures that might point to a better future. He settled on one.

"Why don't you write to the Hendersons and see if there is anything back in Nacmine?"

Chapter Twenty-One
A Legacy of War

Jim looked at the food on his plate. *I'll starve if I stay here much longer.*

He was used to eating family style, with the whole meal set out in serving bowls so that each person could take what they wanted. But at the Hendersons', it was different. Granny, John's mother, was bulimic and Jean found it judicious to portion out the individual plates in the kitchen to prevent any overeating. Jim could not complain. He had been hired on as a fire boss at the Commander and had accepted the invitation from Jean and John to stay with them until he was able to find a place for the family.

Fortunately, within ten days he found a suitable two-bedroom house for sale. A converted store with a large pot-bellied heater in the living room, it was just a block from the mine. John loaned Jim $1,000 interest-free so he could buy it outright. He moved in immediately, batching it — eating well — while waiting for Ida and the children to arrive at the end of June. In the evenings and on weekends, he planted a vegetable garden and also skidded over an old shed, planning to tear it down and build a chicken coop with the lumber. He was excited about another fresh start for himself and his family and, even if he had to be in the mines, he could still raise chickens.

The house had a cold-water tap in the kitchen but no sewer system, so wastewater was collected under the sink in the slop bucket that would soon become one of young Jimmy's responsibilities. The room that had been built as a future bathroom was just large enough to accommodate a single bed and a dresser for Jimmy, who was about to begin high school in the fall. The girls shared one bedroom and Billy had a cot in Ida and Jim's bedroom.

A small creek ran down past the mine and through a ravine between their house and Ayling's store, the post office and the homes of the

Barries, the Hendersons and the Heikkinens. Jimmy quickly developed a friendship with Albie Barrie, whose father, Jack, was the pit boss, and Keith Heikkinen, whose younger sister, Bonnie, hit it off with Cathy. They all preferred running down through the ravine, leaping the creek, and sprinting up the far side as a way of connecting with each other, rather than walking around on the road.

Ida's garden — she had taken it over from Jim when she arrived — flourished and, as a result, she had ample produce for canning. As the various vegetables ripened over the summer, the older children were conscripted to help with preparation: shelling peas, snipping the ends off the beans, slicing the kernels from the corn cobs and massaging the bloody, par-boiled beets to remove their outer layers of skin.

The root cellar kept carrots, turnips, potatoes and onions well into the winter. And, with the mines working full time, Ida could afford to buy peaches and pears to supplement the locally grown crab-apples and Saskatoon berries and her lists of preserves exceeded those of her days on the farm at Coaldale. For a special Christmas treat, she always canned one batch of pears with green food colouring and another with red.

The girls attended school in Nacmine, while Jimmy took the bus to Drumheller for high school. He had an announcement when he returned home on his first day.

"I am now 'Jim'. There were three of us called Jimmy and so our homeroom teacher, Mr. Taylor, suggested we vary it a bit. Jimmy Gibson will keep the name and James Charleston and I will change."

"That may be fine for school, but it will be confusing around here with two Jims," Ida said. "So we will continue to call you 'Jimmy'."

And that was that.

In the winter, the older children would pull a sleigh with a wooden crate on it across the tracks and over to the mine dump to pick coal. The men sorting in the tipple often did not take time to break the 'bone' or rock from the coal, and let those chunks pass to the waste pile along with the slack, the loose coal that remained when the lumps had been screened out. It usually took Jimmy one or two swipes with the back of an axe head to dislodge the coal from the bone. Sometimes other smaller chunks got by the sorters, so the children were able to get two or three loads of good grade coal every Saturday to supplement the supply at home. They had to be careful to avoid 'hot spots' on the pile, as spontaneous combustion often ignited the slack deep within the waste, but in winter it was fun to watch the steam rising and speculate as to the intensity of the inferno beneath.

Because the house was right across the main road from the railway tracks, trains were part of everyday life. There was round-the-clock activity as boxcars full of coal were sorted and connected in preparation for the long journey out of the valley toward markets in the east and west. It did not take long for all members of the family to adjust to the constant noise of shunting boxcars during the night; gondola cars were not used for coal at that time. Empty boxcars, which often came from the west coast where they had delivered grain, were left on the high side of the mine so that they could be moved, one-by-one, into place for filling at the tipple which straddled the tracks at the mine.

Jimmy, along with the other boys, often watched the process of how a miner could set a boxcar in motion with a pry bar that had a sturdy hardwood handle about four-and-a-half feet long attached to a metal apparatus with a hinged lip. He would place the lip on the rail against a wheel and pump the handle up and down. As the wheel slowly began to turn, he pushed the lip forward to retain its position snug against the wheel. The pumping continued until the car had enough momentum to run by itself down the slope. The boys watched carefully as the worker then climbed up the ladder at the end of boxcar and grabbed the brake wheel so that he could stop the rolling car in the right spot in the tipple. After boarding up the open door about halfway, the tipple workers would manoeuvre the loading conveyor into the car and fill it with about eighty tons of coal, shut and seal the door with a metal clip and move it out. The full car then had to be rolled down the grade, again using the pry bar, and stopped as it coupled with other full cars on the siding. During the night, the engine would sort cars and build a train.

On Saturdays, when none of the railway workers were around, Jimmy and his friends would often find a lone car on the sloping siding below the mine and release the brake in order to see if Isaac Newton was right. If, by chance, the car started to roll, they would scamper up the ladders at each end, one person taking charge of the brake wheel and off they would go, making sure they stopped the car before the next switch.

Each working day, the mine whistle blew at four o'clock as the dayshift ended and the afternoon shift began. Then the miners would walk slowly past the house, gradually becoming more erect as they unfolded from eight hours of cramped labour, slowly making the transition from miner to husband and father. At home, one of the treats for the children was the opening of Daddy's black lunch bucket to see if he had left a sandwich — there was something special about a cold jam sandwich sprinkled with coal dust, and the children kept strict mental records as to whose turn it was each day. Billy, who by age three was wearing glasses to correct an eye muscle problem, did not like awaiting his turn and found that if he stood by the gate in the front yard and looked up sweetly at the passing miners, they often responded positively to his touching plea.

"Do you have a samwich?"

Ida was flabbergasted when one of her neighbours told her how cute it was to see Billy out there each day and that she had been putting an extra sandwich in her husband's lunch bucket just for the boy.

The next day, when the shift-change whistle blew, Ida stationed herself at the front window and watched as the miners approached. Sure enough, Billy had moved into position by the front gate and waited as the men came along the road. When the first of the miners neared the bridge, Ida dashed out of the front door.

"Billy, you get in here this minute."

Not waiting for him to react, she scooped him up and had him inside before he knew what hit him.

"I don't want you to ever, ever ask anyone for a sandwich again. Do you understand me? People will think we're poor!"

For some months, Bill, Jim's brother six-years junior, had been writing regularly stating that he, Meg, and their three children wanted to get away from Scotland and its poverty and make a fresh start in Canada. Then in August 1948, he wrote to announce that he had saved enough money for passage to Canada, and they would all be arriving in two months and would need a place to stay.

When Bill and Meg had visited Jim in the hospital in Aldershot and told him how tough life was for them, he had immediately thought back to his own situation years before and his decision to leave Scotland for a better way of life. Meg, a redhead from Fife, did not get along with her Devonside in-laws.

"She is not refined enough," Bill had told Jim when they were alone at one point.

Jim had grown up with a prejudice against Fifers, who were considered lower than their Clackmannanshire counterparts. He had difficulty with her coarse language but had promised to do what he could for his brother. Now, Bill had called him on that promise.

Ida wasn't happy when she read the letter.

"We don't have room for them here," she quickly said.

"I know, and we don't have the money to get them a place."

"It's too bad they won't wait until we get on our feet again."

"They think we're doing well and I'm not going to tell them anything different. I'll think of something."

The answer to the problem lay across the tracks in a little two-room shack beside the ball field. The owner wanted $60 for it.

"It's small," Jim said, "but it will do them until they get on their feet. Do you think we can buy it for them?"

"It will be tight, but we should be able to make do," Ida said. "We won't have to buy many groceries with the garden doing so well."

When it came time to clean out the shack in September, Ida was not feeling well.

"Jim, I'm pregnant again," she announced.

"What did you do that for?" he demanded sharply.

"What's the matter with you?"

"Nothing, why?"

"That's a stupid thing to say."

"Well, we can't afford another child," Jim barked.

"Don't holler at me. I didn't get pregnant by myself! What is the matter with you? You've been snapping at the children and me a lot lately."

"They get on my nerves when they're so loud."

"Are you worried about Bill and Meg coming?"

"No! I'm not worried about anything."

In the next month, they managed to secure some old beds and a table and chairs. With the help of the Hendersons, they were able to get enough bedding, towels and kitchen supplies together for a family of five.

In October, when Bill and Meg arrived with Helen (ten), Mary (six) and James (three), Ida was upset by what she saw.

"Jim, they look like DPs," she said, using the derogatory term for the displaced persons that had flooded into Canada following the war.

"They don't have much, Ida. Do you remember how Pete wrote that he had not seen such poverty when he visited my family during the war?"

"I know, but Bill is the only one who looks clean, and with his long nose and black hair, he looks like he is Jewish."

"What difference does that make? My mother has black hair."

"And Meg looks like she needs a comb and an iron. Did you see her dress? And the kids look like waifs; they're so scrawny and their clothes are too small."

"They've just travelled across the ocean and this whole country. Don't be so harsh."

"We'll have to get them cleaned up and find something for the kids to wear."

There were no extras in the Elliot home, but once again, their more affluent friends were able to provide some cast-off clothing.

Bill and Meg had no money and were entirely dependent on Ida and Jim. They had no awareness of how tough finances were for their Canadian relatives. Fortunately, Bill was quickly hired on at the mine and could begin buying his family's meat and groceries. Meg struggled with Canadian money, customs and quantities and dropped in on Ida almost every day to ask for help in ordering groceries or to get ideas for cooking.

Twice a week, Meg arrived with her laundry wrapped in a sheet, and although Ida had described the process of sorting, Meg could not be bothered to do it and dumped the unsorted laundry into the machine, rinsed it and hung it to dry on Ida's clothesline. When Ida looked at the gray clothes on her line, she was so concerned what others might think that she sorted Meg's next wash and insisted on at least two loads, colours and whites. Meg found it difficult to give up her old ways of doing things and enjoyed the reaction of others to some of her quirkiness, but she came to see that Ida's ideas had merit. After a while, she began to hang out a wash that would not be an embarrassment to her sister-in-law.

The days were wearisome for Ida as she moved through her typical nine months of nausea and felt she had enough to do caring for her own family, which included two preschoolers, without the relentless barrage of questions from Meg and the noise that multiplied profoundly with the addition of three-year-old James. Ida was usually exhausted by the time Meg left for home each afternoon to fry her family's dinner — she loved the black cast-iron pan. Then to add to Ida's fatigue, Bill and Meg dropped in every evening, leaving Helen responsible for her siblings, and stayed quite late. After a couple of weeks, Ida told Jim that she could not take it much longer. The next evening, he responded by bluntly telling Bill and Meg that they had to go home earlier as he and Ida needed their rest. Fortunately, Bill managed the finances in his home and soon had enough resources to buy a larger house at the end of the block near the river. Although they continued to visit Jim and Ida often, they began to develop other friendships.

Jim sold the shack for seventy dollars to one of the Molyneaux boys who lived across the tracks

As the winter passed and moved into the spring of 1949, Jim became more agitated and found fault with everything and everybody around him. He and Ida began arguing about how harsh he had become with the children and how grumpy he was about almost everything in their lives.

He had ordered one hundred leghorn chicks and taught Jimmy and Cathy how to care for them, giving strict orders as to how things should be done — and done right the first time. In the spring, when the chicks were rapidly growing into mature leghorns, one of the chores for the older children was to take a pail and a broom and walk along the railway siding looking for cars that had once carried grain. They swept up whatever kernels they could find and took them home for chicken feed.

After the baby was born on May 2, 1949, Jim mellowed a bit and suggested names for the boy.

"What do you think of Douglas Andrew, Ida? After Matt Douglas and Andy Craig."

Relieved that he was reasonable, she quickly agreed.

That summer, the family hosted a number of visitors from Hillcrest and Lethbridge, as well as the Wilsons, who were now living in East Coulee, where Andy was managing the Atlas mine. Although it meant more work for her, Ida liked having company, as Jim was more civil during their visits. Her brother, Pete, and his English war-bride, Winn, and their four children were among those who visited that summer. It was a cramped, noisy time but everyone found a place to sleep and the children spent most of their days outdoors. One evening, as the adults were sitting in the living room and the older children playing Parcheesi in the kitchen, a passing car backfired. In a flash, Pete was off the chesterfield and onto the floor beside the couch.

As he rose to sit down again, Winn looked at Jim and Ida.

"He still gets spooked by sudden noises," she confided.

Later, in bed, Ida revisited the matter with Jim.

"What was that about?"

"I think Pete's shell-shocked."

In the times when there was no one else around, the arguments between Jim and Ida increased in frequency and intensity until their problems came to a head one day after Ida returned from shopping in Drumheller. John Henderson had come by to see if she needed anything and she had gone with him. Once she'd returned and all the groceries were inside the kitchen door, Jim exploded.

"What's going on, Ida?" he demanded.

"What do you mean?"

"You and John. I've *suspected* for a while."

"For God's sake . . ."

She was relieved that the older children were in school.

"You think I don't know? Sneaking off together. I'll bet *it* happens a lot when I'm on afternoons."

"Jim, stop it!"

She took off her coat and draped it over a kitchen chair.

"How long has this been going on? And he *pretends* to be my friend."

"He *is* your friend and nothing is . . ."

"I've seen how you are turning the children against me . . ."

"Jim, shut up!"

"I know you're trying to get rid of me . . ."

"Will you be quiet for a minute and listen?"

"More lies?"

"Jim, there *is* something wrong."

Her eyes welled with tears.

"So now it's *my* fault! . . . You're the one messing around but it's my fault?"

"No one is messing around, you stupid man. I love you."

"There is nothing wrong with me."

"Either you see a doctor and find out why you are so angry all the time or I am leaving."

"It's you who needs to do something. I don't need to see a doctor! I'm fine!"

"You weren't like this before you joined up."

"So now it's the army's fault."

"Look at what happened to Pete."

"I told you that I'm fine."

Ida began to put the groceries into the cupboard.

"I won't discuss this anymore . . . You decide . . . You have one week."

Jim *did* know something was wrong and suspected that it might be 'war nerves', but he also knew that it only happened to weak men and he certainly was not going to talk about it with Ida. On the other hand, he also knew that Ida always meant what she said, so with great reluctance, he made an appointment to see Doctor Aiello in Drumheller. The doctor assured Jim that what he was experiencing was normal for men coming home from the war and suggested that he see one of the army doctors at the Colonel Belcher Hospital in Calgary. Jim travelled alone by bus to the hospital, where the specialist told him it appeared that his nerves were shot, the result of his overseas experiences. He arranged for him to come back in a month for a more extensive examination and possible treatment. Afterwards, he sent Jim to the DVA office to arrange for transportation costs.

The tension eased a bit at home on Jim's return as he explained to Ida what the doctor had said. She told the children that their father was not well but did not elaborate and suggested that they not upset him too much.

On Jim's next visit, the doctor told him he wanted him to remain in the hospital for a few weeks and that the DVA would cover his salary while he was off work. Jim called on the Hendersons, one of the few families with a telephone, and asked John to let the mine know he would be off, then arranged for a time for Ida to phone him at the hospital.

The doctor suggested that insulin shock treatments would help Jim, as they appeared to be helping people get rid of bad memories.

"You're the doctor," Jim said. "Do what you have to do."

Jim was strapped to a bed and then injected with massive doses of insulin, which put him into a sub-coma for about three hours. Then he was given fruit juice laced with sugar to bring his body back to balance. He would sweat profusely, eat and then drift off to sleep. He endured the horror of this treatment, the infrequent convulsions and ravenous appetite caused by the massive insulin doses because he was determined he would not return to his old ways of thinking. None of this was ever mentioned at home in the presence of the children. All they knew was that dad was sick and had to be in Calgary because he was a veteran.

John and Jean Henderson drove Ida into Calgary to see Jim who, although totally exhausted, had somehow found the beginnings of his lively spirit again. John asked how he was doing.

"Healthy, but dry," he replied.

That fall of 1949, he returned home a changed man, got involved in the management of the Nacmine Dinosaurs baseball team, began fishing again and was easier on the children, often kicking a football back and forth with the little ones.

The next year, when the Hendersons bought a new car, John sold his '33 Chrysler to Jim. It was the same model as King George VI and Queen Elizabeth had ridden in during their 1939 visit to Canada — a great car with hydromatic drive.

The mines operated full-time in that post-war boom period and life for the Elliots was the brightest it had ever been. The children were doing well in school, Jimmy was in the Sea Cadets, Mary Ann and Cathy were both majorettes with the Nacmine school band, and Mary Ann delivered the local *Drumheller Advertiser*. Ida began attending Knox United Church in Drumheller. Jack Barrie sponsored Jim into the Masonic Lodge, where he had an opportunity to meet many of Drumheller's professionals. He spent much of his spare time reading and studying the latest materials on mine rescue and safety procedures and participated in mine rescue competitions. The other miners saw Jim as a fair, hard-working fire boss who did not complain about his work.

Then in the late spring of 1950, Jim began to experience severe pains in his lower back. Each day became torture as he groaned his way out of bed, every movement adding to his distress. Some days just getting dressed took more out of him than a shift digging coal. He continued to work stoically but knew that he could not keep this up.

One August morning as he sat on the edge of the bed, trying to will himself to stand, he turned to Ida.

"I can't do this."

"You've got to see the doctor."

"What if he wants me in the hospital? I have to work."

"You know you can't. We'll manage. We always have."

"I'll do anything to get rid of this pain."

Chapter Twenty-Two
Out of the Pits

Jim heard voices through the confusing fog and could not figure out where he was. Antiseptic smells filled his sensitive nostrils and his stomach spun with nausea. His tongue explored the dry mouth cavity looking for any hint of moisture and his lips felt as though they were stuck together. While he could not move his limbs, he managed to force his eyes open just long enough for the bright light to snap them shut again. Another squint, an involuntary blink — he was on his back looking up at a ceiling, a stark white ceiling.

"You're awake, Mr. Elliot," a disembodied voice gently whispered. "Don't try to move. How are you feeling?"

He rolled his head slowly toward the source and saw a woman dressed to match the ceiling. Choking back the sourness, he managed to slur feebly.

"I'm dry and I think I'm going to be sick."

"It's okay," she said. "I have a basin here. Just relax and let it come."

Cradling his head with one hand, she held the kidney basin close to his face. He relaxed and it came — no time to think about anything except getting rid of the foulness. When the episode was over, she placed the basin on the bedside table, covered it with a small towel and picked up a damp cloth to cool and clean his face. She offered him a straw for a sip of water.

"The doctor will be in shortly. I'm right here if you need me."

Jim closed his eyes to keep out the brightness and tried to feel his feet, but his whole body appeared numb. Prior to his surgery, the hospital staff had explained the procedures they would follow after the operation. The doctor had discovered deteriorating discs in the lower thoracic region, a condition that would require a spinal fusion. Jim had been sent to the Colonel Mewburn Memorial Hospital in Edmonton, where a specialist concurred with the diagnosis and arranged for the surgery. He would be placed on a striker frame bed and have to lie still for the next seven weeks while the bone that was taken from his hip knit to the area on the spinal cord that needed the fusion.

Now the reality of his situation began to seep slowly into his senses. The fog was numbing any pain, but lying flat on what appeared to be more of a board than a bed was very uncomfortable. As his head cleared, the moisture returned to his mouth.

"I could sure use a smoke," he slurred to the nurse.

She smiled and gently patted his shoulder.

"You'll have to wait a bit, I'm afraid."

Jim was a little more alert later when the doctor came to talk about the operation.

"The surgery went well, Mr. Elliot."

"Thanks, Doc."

"You should have no trouble with your recovery as long as you don't try to rush things. You'll do well and there shouldn't be any lasting effects. Just take it easy."

"That's my plan, Doc," Jim said drowsily. "When can I get a smoke?"

"There's a new study that says smoking can slow down healing, especially with bones. This may be a good time to quit."

"Hell, laddie, I think it would be the wrong time to stop smoking."

"Well, once we get you finished with the nausea and you've had a couple of days to start healing, the staff will wheel you into the day room. I'll look in on you tomorrow."

About three hours later, the turning ritual began when two nurses and an orderly came into his room. The three of them continued to chat as they removed the blankets, leaving a lone sheet to protect his modesty. They carefully placed a long padded board, with a space for his face in it, on top of him and began to fasten it with huge straps to the board he was lying on. The whole apparatus was suspended between two large wheels, one at each end of the frame. When everything was tight, one of the nurses grasped a lever and, with a "Here we go. One, two, three," the whole sandwich with Jim as the meat in the middle was rolled over, and what was once up was now down. Jim was now lying on his stomach with a canvas band supporting his forehead. The team then secured each end of the board underneath him to the frame and removed the top board. The orderly knelt down under the frame and held a straw to his lips so he could take a drink of water.

"That's not what I had in mind," Jim said. "But it will do, laddie, thank you."

Jim was more alert the next morning when the doctor made his rounds.

"How are you today? Any pain?"

"I've been better, but it's not too bad."

"It will be uncomfortable for a while as the healing begins. Let the nurse know if you need anything."

"How long before I can get back to work?"

"I'm afraid your days of working in the mine are over."

"What did you say? I've a family to care for."

"Don't worry about anything right now. When we opened you up, we found that the deterioration in the discs was congenital but had certainly been aggravated by your exposure to the stresses of the war. So I am recommending you for that army pension we talked about. It won't be large but once it's through, you'll be well-looked after. I want you to take a year off before you consider your next step. With the pension, you'll get your full salary."

"Then what?"

"We'll have someone from DVA come in and talk to you about future possibilities when you're a bit stronger. In the meantime, I don't want you to worry. Get some rest."

For the next seven weeks, a team came in every three hours and flipped him. He was either lying on his back looking at the ceiling or on his stomach staring at the floor — not much of a view unless one of his friends was visiting and opted to crouch down on the floor and look up at him. However, they mostly saw him face up, as the staff tried to arrange to have him in that position for visiting hours and usually wheeled him into the day room where he could have a smoke. Word spread quickly that he was there and Dunc and Annabelle Heddleston and the Thompsons, including Bunty and her husband Dick, visited him regularly. Bob, Archie and Helen Allen, whom Jim had met when he worked at the Clover Bar mine so long ago, made sure that one of them visited every day. They had never married, choosing to stay with their mother when their father had died years before. They continued in the same house following their mother's death the previous year. Each visitor thought it more appropriate to bring cigarettes than flowers and soon Jim had quite a cache in his bedside table.

Both boards had 'trapdoors' in a convenient spot so he could void his wastes. His back began to heal quickly — with little pain — but the donor site, where they had taken a piece of bone from his hip, took longer and was a constant reminder of the severity of the whole operation. The staff hooked up a rack to hold reading material above or below him so he could read the regular letters from Ida and the children, his family in Scotland, and the friends scattered in mining towns across Alberta. Each Saturday, when the *Edmonton Journal* was delivered, he was also able to keep up-to-date on how things were going in the old country soccer leagues.

Back in Nacmine, Ida maintained a positive outlook, knowing that Jim would return home pain-free. However, she longed to see him as much as he did her. When a letter from Helen Allen, whom she had never met, arrived with the invitation to come and stay with her and her brothers, Ida jumped at the opportunity to go to Edmonton. May Stone, Virgie Barrie's niece, agreed to stay with the children.

As Ida did a last-minute check before leaving, she discovered that the money she had withdrawn from the bank was not in her purse which she had left lying on the bed. Sure that it had been tucked safely into the side pocket, she began a frantic search of the bedroom and noticed the pillows were not the way they had been when she made the bed. With her mind on her loss, she automatically picked up a pillow to fluff it out and set it down properly when she spied all of her money spread out neatly where the pillow had been.

"Doug, what did you do?" she asked her eighteen-month-old toddler.

He could only grin at her and blink his big blue eyes. She had been too busy packing to notice what the boy had been playing with on the bed. Her heart was still racing when the Hendersons pulled up to drive her to the train station in Drumheller.

At the hospital, Ida tentatively approached the bed where Jim lay rigid. He turned his head at the sound of her footstep.

"Come here, lassie. Och, how I've missed you."

"I've missed you, too."

She hesitated as she reached the bed, not sure if she could touch him.

He reached out a hand to pull her close.

"I won't break."

They held each other silently and somewhat awkwardly for a few moments. He was eager to hear all about the children and how things were going back home.

"It's hard being so far away."

"I know, hen, but it won't be too much longer. How are the bairns?"

"They send their love as does half of Nacmine. You look a little pale. Are you doing okay?"

He rubbed his hand over his chin.

"It's just that I don't have to scrub my face raw to get the coal off anymore. Get the nurse to wheel me to the day room, will you? I need a smoke."

Facial contact and handholding had to be the extent of their intimacy. They talked briefly about their future but agreed to wait until he was back home before making any major decisions.

Jim told her that the hospital was named after a man who got his start as a doctor at the Galt mine in Lethbridge.

"It is kind of strange, you ken, I've worked there," he said, "and now it's here that they do the operation that takes me out of the mines."

"Maybe it's the way things are supposed to be," Ida replied, squeezing his hand.

Ida had an opportunity to see some of Edmonton during that week as the Allens took a different route home from the hospital each day. She was enthralled by the size of the city as it sprawled out on both sides of the North Saskatchewan River. When it came time for her to return to Nacmine, it was hard leaving Jim, but she was relieved that he was doing so well and in such capable hands. More than once during her visit, one of the staff had remarked what a great patient he was.

The day Jim's ordeal on the striker frame ended, he was taken to the plaster room where his body was encased in a plaster cast from neck to hips.

"This will keep you from bending your back and putting any stress on those discs," the doctor advised.

"How long will this take?" Jim asked, thumping the side of his cast.

"About four months, just to make sure."

"Four months? I can't stay here for that long."

"Once we get you used to it and you are able to walk on your own, we'll send you back to that beautiful wife of yours."

The next morning, when Jim awoke, he rang for the orderly.

"I've had enough with bottles and bedpans — I need to get to the toilet."

The orderly helped him into an upright position, but as Jim tried to put his weight onto his feet, he crumpled into the man's arms — the result of the lack of muscle use, the weight of the cast and the dizziness that engulfed him.

"That's a bugger!" Jim exclaimed.

"A little at a time and we'll make it," the orderly encouraged.

Jim's strong determination to get home quickly drove him to gain the strength needed to take his body from the horizontal to the vertical. Soon, he was onto his feet and across the room with no assistance. He was discharged in mid-December, after eight weeks in Mewburn. Bob Allen drove him to the train station and asked the conductor to make sure his friend was comfortable on the trip.

That Christmas was one of the most joyous the Elliot family ever had. Ida and the five children, all with moist eyes, sat quietly as Jim began to play *Away in a Manger* on his mouth organ, then they all joined in and sang the song that had become the family Christmas anthem. Jim was in great spirits on Hogmanay as he welcomed the first-footers, who each made sure that the Elliots were on their route.

"What a way to celebrate my forty-third birthday," Jim said as he hoisted a wee dram in the early morning hours of January 1, 1951.

While Jim could sit upright, it was still very tiring. As a result, he spent much of his time lying prone on the chesterfield in the living room, with a number of cushions supporting him. As it was too treacherous for him to walk on the icy roads, his outdoor activity was limited to his freezing, shuffling trips to the outhouse. Jimmy, in preparation for his father's return, had carefully tended the path by emptying the ashes from the stove and heater along it, giving Jim a solid walkway with plenty of traction. He read a great deal, but with limited movement he was uncomfortable much of the time. The itching skin under the cast was unbearable, and he would often ask Ida to hand him a knitting needle so he could poke it between the cast and his body and scratch the irritating parts.

Fortunately, the cast was removed in April before the weather got too warm. Jim began to walk to the post office daily in order to strengthen his muscles.

The children continued to grow and mature through all of this. In June, seventeen-year-old Jimmy graduated from high school and, through a connection with one of Jim's Masonic brothers, Blythe Davidson, the manager of Canadian Utilities, was soon hired as an office assistant for the power company. Cathy had just finished her first year of high school and she and Mary Ann went to Calgary with the Nacmine band to march in the Stampede parade in the first week of July. Jimmy was also in the parade as part of the colour party with the local sea cadet group. Billy was a quiet boy and did not demand much attention, while Doug made up for the both of them as he moved through his terrible twos. The Hendersons continued to be generous and made their piano available to Cathy when she began lessons, and, since they had indoor plumbing, they invited Cathy and

Mary Ann to bathe at their house. Jimmy showered at the mine washhouse each weekend with the other teenage boys and the two youngest continued to bathe in the square washtub set out in the middle of the kitchen.

As Jim became stronger, he got restless and Ida suggested that it was time to make decisions about their future. One evening in late July, they talked about what to do next.

"I've been doing a lot of thinking as I walk around this town. I want to get out. There is nothing for me here and I don't want to be a lamp man or do anything else around the tipple. What would you say about leaving the valley and going to Edmonton?"

"What will you do there?"

"I don't know. But while I was in the hospital, I saw lots of jobs advertised in the *Journal* and the DVA helps veterans find work. Archie and Bob offered to keep a lookout for me."

"But the big city kind of scares me, Jim. And what about the children? Jimmy's got a good job and I hate to move the girls again."

"They'll do fine," he continued with enthusiasm. "Jimmy does have a secure job and has signed up for a correspondence course in accounting from Queens University. He says that he can live with one of the people who works with him. Ida, there will be more opportunities in the city for the other children."

"What about a house?"

"We'll find something. I'll look around next week when I'm in Edmonton for my check-up."

In Edmonton, Bob and Archie Allen drove Jim to look at a number of places and he found a nice little house near the Allens, in the south end of the city. Since it was more expensive than the one he owned in Nacmine, when he returned home, he arranged another loan from John Henderson to make up the difference. So in the early fall of 1951, the family was on the move again, most of them to 9636 – 76th Avenue in south Edmonton.

Jimmy stayed behind to continue on at his job. He went to live with the Bart and Wendy Howard and their four-year-old daughter, Cindy. Bart was the sales manager in the power company office and their home had a good-sized second-floor room for Jimmy.

Since Jim could not drive the car yet, he had arranged for a young miner to drive it up on a future weekend. As the train pulled out of the valley, Ida asked Cathy to keep an eye on the boys as Doug was trying to follow Billy on his exploration of the rail coach. Cathy was relieved to have something to do and hide the sadness that had been part of her ever since the move was announced. At sixteen, she would have to face a new city, a new school and no friends.

"Don't worry, Mom, I'll take them for a walk into the next car," she said over her shoulder.

The boys were delighted to have a little freedom. Mary Ann, at twelve, saw it all as an adventure. *Imagine going to a big city to live!*

Ida and Jim each bolstered the other with encouraging words.

"We know a lot of people in Edmonton," Jim said. "At least we won't have to worry about making new friends."

Bob Allan was at the Whyte Avenue Station when the train pulled into south Edmonton and soon the newcomers were enjoying the hospitality of this generous Tillicoultry family.

Once they were settled in their new home and the girls were registered in school, Jim began to look for work. One of the first places he went was the CPR, where he had once worked for a month, but he was told that, at forty-three years of age, he was too old. The experience brought back memories of his desperate searching for work in this same city so many years earlier. Then, he had been on his own. Now he had a wife and four children to take care of.

Each day, he read the want ads and rode buses all over the city to follow up, only to return home, jobless and more discouraged than he had been the previous day. He was in the prime of his life and no one wanted him. *Will it never end? Just one month left of DVA benefits. I won't go on relief!*

Then one day, an ad that began with "Are you a veteran?" grabbed his attention. A construction firm, Brown and Root, was looking for security guards during construction of the new Canadian Celanese Corporation plant, known as Chemcell. It took Jim almost two hours by bus to find his way to the construction site across the North Saskatchewan River, just east of the city. Following a short interview with a very understanding personnel officer, he was immediately hired, given a uniform, and assigned to the front gate that was used by staff.

However, the new plant was northeast of the city and he had bought a house in the southwest. Gas was too expensive for him to commute by car so he had to use the public transit system. It took over an hour to connect with the Brown and Root company bus that ran along 118th Avenue and another half-hour from there to the work site. He had hardly begun work when winter set in and the three hours on the bus each day were an added burden to the low-paying job. His salary was just $46 per week and did not go very far with four children at home and a mortgage to pay, but it was work and he would stick with it.

All of the other security guards were veterans of World War I and at least twenty years his senior. They appeared happy with their assignments of general patrol duty around the site and staffing the other gates used by suppliers and contractors. During breaks, they competed for the floor, each trying to outdo the others with war stories. Jim, who did not want to go there, kept up his end with humorous anecdotes from his Scottish days or else tuned them out as he thought about his future. *Is this my lot? Will I ever make anything of myself?*

"We might as well sell the car," he told Ida one day. "I can't afford to drive it and the insurance is coming up next month. We could use the money."

"Would there be enough to buy a fridge? We'd manage so much better if we could keep food longer."

"If that's what you want, then get it," Jim replied firmly. "Jimmy said he could buy appliances wholesale through Canadian Utilities. Write to him."

Jimmy made all the arrangements through Bart, who had the fridge shipped from the head office in Edmonton. Ida had her first refrigerator, a Leonard.

In the spring of 1952, Jim had reached the point where he either had to quit his job or move closer to the plant. He and Ida decided on the latter and soon found a three-bedroom house on 67th Street in the Highlands district of northeast Edmonton, about two blocks from the bus route on 118th Avenue. Once they decided to relocate, Cathy balked.

"I won't go!" she announced. "It's too hard changing schools all the time."

"You have no choice, young lady," Ida snapped back at her. "We're moving!"

"Maybe you are, but I'm not. I'll stay with one of my friends."

"You'll do nothing of the sort," Ida said firmly.

"It's not fair. You do whatever you want and never think about us kids. You're mean."

"Don't talk to your mother like that," Jim interjected. "We don't have a choice. I need to work and the work is on the other side of the city."

"It's not so bad taking a bus across the city. I don't see why you can't keep on."

"Cathy!" Ida said emphatically. "Your father works very hard to keep this family together and you are the one who is not being fair."

"Nobody thinks about me." She stopped to blow her nose then continued, "It's stupid to leave here when I have only three more months of grade eleven."

"Come on, hen," Jim said tenderly to his daughter, "we have to move and we are not leaving you behind."

"Well then, can *I* take the bus every day?" she sobbed.

"If you think you can do it, then give it a try," Jim agreed, "I promise that you won't ever have to change schools again."

They reached a compromise and Cathy became the one who did the long bus trips to remain at Strathcona until the term ended in June.

When the administration building was completed in late 1952, the Chemcell management team moved in. Jim greeted the new faces with his typical 'Hello, laddie' or 'lassie' as they moved past his post at the gate, and some of them stopped for a bit of conversation with the man they began to call 'Scotty'. Dick Toews, the personnel manager, a former RCMP officer, was especially drawn to this sentry at the gate and, over a few weeks, got to know a little about this cheery Scot.

"What did you do before you came here?" he asked one morning.

"I was a fire boss in the mines."

"I don't know what that is."

"I was responsible for the blasting and the safety of the crew."

"You have first aid, then?"

"Aye, I've even taught it."

"Scotty, can you get a break sometime and come up to my office? I have an idea I want to run by you."

"Aye, I'll do that today."

When Jim arrived in his office, Dick began to explain.

"Plant safety is a priority in Chemcell. We have been instructed to develop a state-of-the-art first-aid station to accommodate the eight hundred employees we will have onsite."

"What are you thinking, laddie?"

"We have a doctor lined up to come in part-time and will hire three nurses who will rotate in shifts. But I would really like to have a first-aid man here full-time, and you impress me as the kind of person I want."

"You mean you'd hire me?"

"I think you could do a great job for us."

"I can learn quickly."

"We'll send you off to Arnprior, Ontario, to attend an up-to-date course in industrial first-aid before the plant opens."

"When do I start?"

"As soon as Brown and Root can get by without you."

In the spring of 1953, Jim tendered his resignation and became one of the first employees of Chemcell. The increase in salary took a great deal of pressure off the family, especially Jim who savoured his good fortune; a new job, working indoors and doing something he had always dreamed of.

He assisted a consultant in industrial first-aid who had been hired by the company to set up the first-aid and safety division. Jim happily unpacked supplies and sorted them onto shelves and into cupboards, set out new equipment and familiarized himself with the components of his new work environment. Two weeks at the federal government training centre in Arnprior reinforced much of what he already knew about first-aid and provided him with a solid base to understand and work in a new milieu.

Prior to the plant opening, Jim was kept busy treating cuts, scrapes and other minor injuries sustained by the growing staff as they familiarized themselves with their new equipment and work environment. He quickly became quite adept at rinsing eyes that had been exposed to noxious fumes, bandaging fingers that came into contact with steel strapping in the cellulose boxing unit and straightening the safety glasses that were mandatory in the workplace. Some of the men who smoked found that if they 'accidentally' dropped their glasses onto a conveyor belt as a box was being loaded, they could get a break from their no-smoking work environment when the supervisor sent them to see Scotty in first-aid. Although a smoker himself, Jim took it upon himself to police this activity and told the offenders that he was going to charge them for each pair of new glasses.

He was happier than Ida had ever seen him.

"Ida, for the first time in my life, I can go to work and not get filthy with coal dust or farm dust or be standing outside in the cold."

"Well, life just gets better every day. The kids are happier, too."

"Now we can start doing some of the things we've never been able to do."

"As long as you're happy, I don't need much else."

By the time the plant opened that summer and went into production, Dick Toews had hired three young nurses to work in the first-aid department, Marie Davidson, Lauga Eggilsson and Marion Mullard, all of whom quickly grew to love Scotty, as he did them. Dr. Russ Taylor, a family physician, was contracted to visit the plant once a week and he, too, appreciated Jim's talents and common-sense intuitiveness with first-aid.

With their newest and most modern plant into full production, the management of the Celanese Corporation of America decided to show off their newest Canadian facilities by hosting an open house in August 1953. The Allens picked Ida up on their way to the event. It was quite a day with many displays, tours and refreshments. When Ida and her friends reached the first-aid department, they were greeted by three young and attractive uniformed nurses and a very handsome forty-four-year-old man.

"My God, Jim," Ida said, seeing him in his work clothes for the first time. "You look like a doctor."

"Yes, lassie," he beamed. "If my Aunt Liz could see me now!"

Scotty's whiter-than-white coat showed how far behind he had left the belly of blackness down in the mines.

Epilogue

Scotty stayed at Chemcell for the next twenty years. During that time most of the eight hundred employees had one reason or another to encounter this well-loved Scotsman. He continued to teach first-aid and, with his friend, Jimmy Thompson, revised the St. John Ambulance first-aid manual. In recognition of this, he was presented with a medal by the Governor General of Canada on behalf of the Order of St. John.

During this time, Ida began to work outside of her home for the first time, when she accepted a position in the cafeteria at Sears. She loved food preparation and soon became a mentor for many of the young staff.

Ida and Jim developed many friendships in the Chemcell community and spent part of each summer holidaying with them as they explored western Canada in their recreation vehicles.

They retired to Duncan, BC, in 1973, where they both became involved in the seniors centre, Jim soon becoming president. Ida learned to drive after she turned sixty and took on an Avon route. Both of them were made life members of the Duncan Seniors Centre. Ida was also awarded life memberships in the Order of the Eastern Star and the United Church Women.

Ida and Jim continued to enjoy their RV's, finally buying a motorhome for their later years of travel. They also travelled to Europe on numerous occasions.

They were proud of their children and their accomplishments. Jimmy, who became Jim again when he went to university, was ordained as a United Church minister. Cathy worked in office support jobs until she married an accountant, spending the rest of her time in the United States. Mary Ann became a nurse and was incredibly proud when her father instructed her class in first-aid during her training at the Edmonton General Hospital.

She married a United Church minister and continued to live in Edmonton. Bill studied education at the University of Alberta, taught and then became a principal in Wetaskiwin, where he went on to be a city councilor and then mayor. Doug studied accounting and moved to Whitehorse.

Jim died in his 90th year and Ida in her 100th.

In addition to their five children and their spouses, they are survived by fourteen grandchildren and seventeen great-grandchildren.

Acknowledgements

This book began as the opening paragraph of a personal history to share with my grandchildren. That changed immensely when I joined one of the writing groups mentored by the much-published author and teacher, Betty Keller — a gift to writers and to writing. I serendipitously found myself in a group with Rebecca Hendry and Vici Johnstone. Experiencing their creative writing and the gentle critique of my work pushed me farther and farther back into my history and also helped me develop a more inventive way to express the story until that paragraph became this book. I am deeply indebted to them for getting me underway and for the others who followed in Betty's groups over the next five years.

My parents were storytellers and, while Dad hogged the stage when he was alive, Mom took over admirably for the next ten years until her memory began to fail. I am grateful for the stories lived and told, and for the audiotape Dad made of his life in Scotland and early days in Canada.

I have tried to be true to the historic details of the story and take responsibility for writing what I have heard and experienced. Some of the conversations are direct quotes from tapes, others from memory and still others from the recesses of my mind and what I believe to be genetic memory.

My siblings, especially MaryAnn (as she now signs her name) and her memory of Nacmine days, added to the stories. My cousin Nadine's notes on her interviews with Mom added much to my resources.

My children have been encouraging through the process and my eldest grandchild, James (what other name could he have?), always greets me with, "Hi, Grandpa, how are you doing? How's the book coming?"

Through it all, my wife Geniene has been a constant rock and support, loving me through the dry times and encouraging me to keep going when the energy is there.

Jo Blackmore of Granville Island Publishing has been patient to the extreme. My editor Kyle Hawke, with his attention to detail and creative guidance, encouraged me to keep at it.

Thank you to all who have helped make this memoir a reality.

**Jim Elliot** was born in the coal-mining town of Drinnan, Alberta. Over the next eighteen years of his life, the family would live in most mining areas of Alberta, going from Mountain Park to Lethbridge, to the Drumheller Valley, to the Crowsnest Pass, and then to the Okanagan while Jim's father was overseas during World War II. After the war it was back to the Crowsnest Pass and then to the Drumheller valley again. Jim attended eleven schools in twelve years.

Following high school in Drumheller, Jim began working in the accounting office of Canadian Utilities, while studying accounting by correspondence. After three years of this he felt 'called' into the ministry of the United Church and moved to Edmonton, where he obtained a BA from the University of Alberta and a divinity degree from St. Stephen's College.

Upon ordination he was sent to Magrath in southern Alberta and, after three years there, moved to Edmonton. Five years later he was called to St. David's, the largest United Church in Calgary, where he served for seven years. This was followed by pastorates in Richmond and then Highlands in North Vancouver. After twenty-four years in suburban ministry, Jim felt called into outreach work and spent the next three years working with the Gitxsan First Nations people in the Hazelton area. This work led him to accept the invitation to direct the mission at First United Church in the downtown eastside of Vancouver. The ten years at First United were the most profound of his ministry.

He and his spouse, Geniene, retired to Halfmoon Bay on BC's Sunshine Coast, where Jim volunteered with the local hospice society and joined a writing group that encouraged him to do this family history.

Jim gets his energy from the sun, the sea and the joy of ten grandchildren.